I0022928

ANCIENT WISDOM *for* MODERN CAMPAIGNS

Lessons from:
SUN TZU'S **ART OF WAR**

CAITLIN HUXLEY

ANCIENT WISDOM FOR MODERN CAMPAIGNS

Copyright © 2023 by Caitlin Huxley

The Art of War is in the public domain. All original additions are copyright © 2023 by Caitlin Huxley and may not be reproduced in any form without written permission from the publisher or author, except as permitted by U.S. copyright law.

This publication is designed to provide accurate and authoritative information in regard to the subject matter covered. It is sold with the understanding that neither the author nor the publisher is engaged in rendering legal, investment, accounting or other professional services. The advice and strategies contained herein may not be suitable for your situation. You should consult with a professional when appropriate. Neither the publisher nor the author shall be liable for any loss of profit or any other commercial damages, including but not limited to special, incidental, consequential, personal, or other damages.

For information contact :

Caitlin Huxley

Huxley Strategies, LLC

822 Guilford Avenue, 1535

Baltimore, MD - 21202

www.HuxleyStrategies.com

Library of Congress Control Number: 2023917990

Paperback ISBN:	979-8-9890737-0-2
E-book ISBN:	979-8-9890737-1-9

First Edition: October 2023

10 9 8 7 6 5 4 3 2 1

Find Free Guides and Downloadable Resources at:

www.HuxleyStrategies.com

The Campaign Managers Toolbox is designed to provide new candidates, campaign staff, and volunteers with the skills they need to succeed in the political arena. These guides cover essential topics such as organization building, fundraising, voter contact, GOTV (get out the vote), and more.

Designed for both newcomers and seasoned veterans, these guides and templates will set your campaign on the path to victory and help you navigate the complexities of modern politics

CONTENTS

Introduction

I believe that Sun Tzu's Art of War holds timeless truths that can be applied to many situations in modern times, including to the difficult landscape of political campaigns. The art of strategy, the importance of adaptability, and understanding one's environment are as critical to a modern political campaign as they were on the battlefield when Sun Tzu wrote his manual.

The best time to read this book is before and during the campaign planning process; as you do your research, sketch out your plan, and imagine your campaign developing during the different phases of the campaign cycle. This is when your mind will be in accordance with the nature of your election, and this is when you will get the most out of the thought exercises within.

As you read through this book, you should perform Premeditatio Malorum (the Premeditation of Evils), a Stoic practice in which one visualizes potential difficulties in advance. By trying to anticipate challenges now you can prepare for them in your mind long before you encounter them. Then, in the heat of the moment, you will be capable of turning these challenges into opportunities.

By embracing this ancient wisdom and engaging in thoughtful anticipation, you can transform yourself from a simple campaigner into a strategic general, ready to navigate any political situation with confidence and resilience.

This book is organized into six parts, each focusing on an aspect of the Art of War and several of Sun Tzu's original chapters. Some of the larger chapters are then further divided into sections.

Part 1: Campaign Planning, comprises chapters on drafting your campaign plan and understanding the costs associated with waging political warfare.

Part 2: Strategy, focuses on the importance of tailoring your strategy to the circumstances, estimating your chances for achieving victory, and best practices for avoiding common pitfalls. Here we will also explore how your chosen strategy will define your tactics and how to choose among the possible tactics available to you to maximize your chances of success.

In **Part 3: Management Principles**, we delve into the process of building and managing your staff and volunteers who are the lifeblood of your campaign, and the right way to grow your organization strategically in order to reach your goals.

Part 4: Adapting your Strategy, covers the essentials of analysing strengths and weaknesses, making contingency plans, seizing opportunities, and risk management. This part emphasizes the importance of being flexible and adaptable to changing circumstances.

Part 5: Navigating the Political Landscape, provides insights into positioning both on issues and physically within your district. We will discuss understanding the dispositions unique to your district and responding to circumstances that exist within.

Finally, **Part 6: Scandal & Spies**, explores some of the more contentious aspects of campaigning: scandals, negative messaging, and intelligence gathering. While some may find these topics uncomfortable or contrary to the way they want to run their race, remember that your opponents may not share these reservations and therefore these chapters highlight the need to comprehend and prepare defenses against such strategies.

As we journey through each aspect of the campaign, we will explore the ancient wisdom of Sun Tzu and I will do my best to add in the lessons I have learned from my personal experiences. As you read through the book, I encourage you to pause regularly to reflect on the advice, and specifics of your race. Through this exploration, I hope to prepare you for the campaign trail and equip you with the tools you will need to succeed on election day.

A Note on Generalship

The general who advances without coveting fame and retreats without fearing disgrace, whose only thought is to protect his country and do good service for his sovereign, is the jewel of the kingdom.

The best campaign manager is one who is willing to advance without desire for personal glory, and is willing to retreat without worrying about damaging their reputation. This book is written for candidates and campaign managers who see themselves as contributors to a cause larger than themselves. Remember that the lessons you learn from your experiences on your campaigns, especially the challenging ones, are priceless. When looking back at my own past campaigns, and being faced with my own errors, I try to remind myself that every unfortunate result is a relatively inexpensive lesson to learn in the grand scheme of things, and will only serve to make me a better general.

Presumably, you got involved in politics in the first place because you cared about something bigger than yourself and your political career. While there is money to be made in campaigning, at the end of the day, your salary comes from donations to what is essentially a non-profit organization. Similarly, there is plenty of glory to be gained, but if you allow the pursuit of glory to become your primary objective you'll begin to see campaigns only in terms of personal gain, rather than as service to the people of the district. Over time you may experience a sense of disillusionment and become jaded. This is quite common among professional staffers and politicians. But, if you find that you no longer have the burning embers that once ignited your passion, you may be better off in the private sector where most staffers eventually find their place.

This is a call to remember your initial motivations, to act in the best interest of your party and your ideals. It's an encouragement to support candidates whose values align with yours, and a caution against supporting those who you know have selfish motivations.

The student of war who is unversed in the art of varying his plans will fail to make the best use of his men. Hence in the wise leader's plans, considerations of advantage and of disadvantage will be blended together.

If our expectation of advantage is tempered in this way, we may succeed in accomplishing the essential part of our schemes. If, in the midst of difficulties, we are always ready to seize an advantage, we may extricate ourselves from misfortune.

Holding the advantage, a key concept in political campaigns and military strategy, is a cornerstone of victory. In this context, advantage is derived from the Clausewitzian concept of achieving a position of superior strength or influence over one's adversaries. It's not merely about having more resources or better assets, but rather the ability to maximize the strategic impact of those you already have. In political campaigns, this can translate to an effective messaging strategy, a robust grassroots organization, key endorsements, or favorable public opinion. Recognizing, cultivating, and leveraging such advantages can tip the balance of a campaign or conflict in one's favor.

If you can wrest the advantage from every situation, volunteers will flock to your side, donations will roll in, and the largest voting blocs will swing in your favor. But if your opponent has the advantage, you'll be at his mercy. Everything falls apart, and it can feel like there's no way to win. To hold the advantage is to hold the keys to success.

Campaigning, much like war, is an art form that requires both knowledge and adaptability. As we delve into the heart of this book, the reader is invited to become a student of all past campaigns. After your election is over, and you look forward to the next cycle, you should try to stay up-to-date on the latest training courses, webinars, and books about campaigning. By studying the tactics that have worked for others in the past, and understanding how those tactics were eventually defeated, you can gain valuable insight into how to improve the way you run your own campaigns.

This is the only way you will succeed. If you are vigilantly looking for opportunities to derive an advantage for your campaign, you can side-step nearly any disadvantage, or fix any harmful situation. A truly effective campaign manager must be able to balance the pursuit of advantage with the need to avoid disadvantage. By carefully weighing the pros and cons of each potential move, you can avoid making rash decisions that may ultimately prove costly.

Part 1: Campaign Planning

"In preparing for battle I have always found that plans are useless but planning is indispensable."

Dwight D. Eisenhower

This sentiment holds as true in politics as it does in war. Having a plan allows you to decide what you will do if everything goes perfect. A plan serves as your guiding star, plotting the course from your current position to your desired destination.

Of course, no plan unfolds flawlessly. Circumstances will invariably evolve and unexpected challenges will arise. Regardless, having a well-structured plan equips you to adapt, providing a foundation from which you can pivot and adjust.

In Part 1, "Campaign Planning," we delve into the process of drafting a campaign plan, drawing parallels to the strategic principles outlined by Sun Tzu. These initial chapters underscore the idea that the art of campaigning, much like the art of war, is vital to the State. As such it warrants careful study and preparation.

This section lays the groundwork for your campaign journey, emphasizing that planning is not merely a step to take before the action of the campaign begins, but an iterative process that will continue to evolve as you navigate the road ahead.

I. Laying Plans

The art of war is of vital importance to the State. It is a matter of life and death, a road either to safety or to ruin. Hence it is a subject of inquiry which can on no account be neglected.

The art of campaigning is full of intrigue and deception. The ability to persuade voters is crucial for those seeking power and influence. Just as a state must maintain a military and constantly train and prepare them for battle, even in times of peace, political players must constantly hone their campaign skills. This is a subject worthy of serious study and attention. Whether you're a candidate seeking office or a supporter of a political movement, understanding the art of campaigning is vital to your success.

For your campaign to succeed, you will need to put in the time and effort to understand your district and develop a strategy that will work for you. Without this focus and dedication, your chances of winning are little more than a roll of the dice. As Sun Tzu said, it is a subject of inquiry which can on no account be neglected.

It's important to remember that Sun Tzu was writing about the art of warfare, not the science of it. Like art, a successful campaign strategy requires creativity, intuition, and the ability to think outside the box. Sun Tzu's lists are starting points from which to view your campaign. As a campaign leader, you are responsible for molding and modifying these principles to fit your unique situation. These lists are not exhaustive, and you should not feel surprised if you think of something that Sun Tzu doesn't mention.

> *The art of war, then, is governed by five constant factors, to be taken into account in one's deliberations, when seeking to determine the conditions obtaining in the field. These are:*
> *(1) The Moral Law, (2) Heaven, (3) Earth,*
> *(4) The Commander, and (5) Method and discipline.*
> *These five factors should be familiar to every general: he who knows them will be victorious; he who knows them not will fail.*

In the art of war, five factors determine victory or defeat, and you should carefully consider each. Imagine potential problems and how you might address them. Only then can you develop a plan that gives your campaign the best chance of victory. Ignore these, and you may find yourself on the losing side on election day.

This chapter poses questions which might be hard to answer if this is your first campaign; this is why it is so important to surround yourself with experienced advisors and staff. Opposition research can also help, but that will be discussed later.

The Moral Law causes the people to be in complete accord with their ruler so that they will follow him regardless of their lives, undismayed by any danger.

The first campaign I managed was a traditional Republican candidate on the Northside of Chicago. I didn't know it yet, but we had no chance of winning, and as a result, every step of the campaign was a struggle. To fill the gap, we hired interns for the traditional staffer roles, and we made more voter contacts than any other State Representative or Senator that year. But it didn't matter. Despite our efforts, we lacked the possession of the Moral Law, and our campaign was doomed from the start.

If you're facing scandal or your opponent is attacking you, the Moral Law is what keeps people on your side.

How much do each candidate's policies overlap with those held by the majority of likely voters in the district? In order to understand the political landscape of a district, you need to examine the overlap between each candidate's policies and the views of likely voters. This includes considering the dominant party in the area, the top issues on voters' minds, the candidates' own demographics, and how well they fit the district. By repeating this analysis for each candidate, you can better understand how well-aligned they are with the electorate.

Who among the candidates is most in touch with the district and its people? Who more closely resembles the people who live and vote here? Who supports the policies that the actual voters in the district support? Rely on polls here, as this is less likely to be biased than word of mouth or "what the doors are saying."

Heaven signifies night and day, cold and heat, times and seasons.

The dynamics of the political landscape are much like the shifting seasons; understanding their rhythm can mean the difference between victory and defeat.

In politics, this is how much time will be available for campaigning between now and election day and what the weather will be like while you're knocking on doors. The campaign that starts earliest will have the most flexibility on what to do and when, and if your team has the time to adapt and respond to new developments you will be more likely to succeed.

For example, in Illinois, the window for collecting petition signatures to get on the ballot is from fall to early winter, and often the snow is falling, and the winds are howling. Not exactly prime time for political campaigning, but you have no other choice. I have seen plenty of would-be candidates wait too long, and they end up having to collect their signatures all alone in the bitter cold. Many never make it on the ballot because of this easily avoidable mistake.

As Sun Tzu noted, natural elements can play a significant role in the success or failure of a campaign. Despite this, there is no concrete answer about the best time to start or how long you should spend gathering allies and resources. Before you start, you will need to evaluate the specifics of your race.

Earth comprises distances, great and small; danger and security; open ground and narrow passes; the chances of life and death.

Mastering your district's unique demographics and geography can be the game-changer, turning an ordinary campaign into an extraordinary triumph. Consider the size and population distribution of your district. By mapping out your districts' demographics in this way, you can begin to sketch out a voter targeting plan that meets your campaigns unique needs.

For example, focusing on friendly areas in the early stages of the campaign, allows you to collect petitions and recruit volunteers without spending excessive time in opposition territory early on. With today's technology allowing you to reach supporter groups no matter where they live (often called micro-targeting), this approach may not be as critical as it once was, but when it comes to maximizing campaign efforts, every little bit helps. This is using Earth to your advantage.

Don't underestimate the importance of Earth in shaping your campaign plan, especially if you're a new candidate! Whether you're running for city council or for president, you've got to get out there and knock on doors, attend events, and be visible. It can be tough, especially if you're in a big district, where the geography is wildly different from place to place. To run a successful campaign, you need to fully understand your constituents' concerns and how they differ from neighborhood to neighborhood and from town to town.

The Commander *stands for the virtues of wisdom, sincerity, benevolence, courage, and strictness.*

It might seem obvious, but a successful campaign is not just about the candidate but also the team behind them. Think about each candidate's Campaign Manager and their close advisors. Which group is more professional? Which has the most relevant experience to the situations specific to your race? Which side maintains the highest quality relationships with volunteers, donors, and voters?

Your Campaign Manager and advisors are the backbone of your campaign, and their experience and relationships can make or break your chances. It is said that you are the average of the five people you spend the most time with. Thus, you should carefully consider the people involved with your campaign.

Further, filling your campaign with responsible, experienced politicos will provide a necessary balance to the more eccentric or out-of-touch individuals occasionally found on campaigns. Those sign-waving parties and lit drops might seem fun and time efficient, but they don't do anything to deliver your message to actual voters. Beware the people who would have you prioritize things which cost more than they achieve.

Who has the most experienced and connected campaign advisors? If you discover the answer to that question isn't you, think about what you can do to make up the difference. There are plenty of classes and books you can provide to your staff to try to overcome the disadvantage.

> By **Method and Discipline** are to be understood the marshaling of the army in its proper subdivisions, the graduations of rank among the officers, the maintenance of roads by which supplies may reach the army, and the control of military expenditure.

Pay special attention to how your campaign staff are trained and managed and to the quality of your communications with them. A well-run campaign will have a clear hierarchy and division of labor, with a professional manager overseeing specialized staff in each department. This will lead to order & success while, a haphazard group of volunteers lacking direction and coordination can lead to chaos and failure.

When it comes to leading a campaign, experience matters. But even a seasoned commander can only win with a solid team behind them. That's why I make it a point to train all my staff thoroughly; I hire experienced managers, fly in training organizations like the Leadership Institute, and have weekly meetings to cover best practices relevant to the next steps and phases of the campaign as they approach. Then I equip my staff to pass on their knowledge to their volunteers and interns. People overlook this often, especially incumbents who have grown slothful after years of easy victories.

Ask yourself which campaign is more professionally managed and better disciplined? If the answer is your opponent, think how you can tip the scales in your favor. Focusing on how you marshal your team is one of the easiest ways to gain an advantage over your opponent.

* * *

Feeling a bit fatigued and overwhelmed during this part of the planning process is not unusual. Rest assured that you don't need to score higher than your opponent in every aspect of your campaign or hold every single advantage available. But every little bit helps, so you should seek out every opportunity you identify and try to exploit it as best you can. Small advantages, when combined, can add up and make a big difference.

According as circumstances are favorable, one should modify one's plans.

Having a well-thought-out plan is probably the most important factor in determining success. Be as diligent as possible, ask yourself the questions above, and follow the advice that follows. Consider anything that might give you an advantage in your campaign, and if over the course of the campaign, you believe some newly discovered strategy will make you more likely to win the election, change your plans to incorporate it.

All warfare is based on deception.

When your opponent doesn't know what you're planning, they can't do anything to stop you. That's why a key to success in warfare is the ability to deceive your enemy. So, how do you pull off a good deception? Simple, it all boils down to perception. If you can control your opponent's perception, you can manipulate their actions and reactions. This means using clever tactics, disguising your intentions, and even maybe a little psychological warfare to throw your opponent off balance.

Many candidates are hesitant to start building their organization or reaching out to voters for fear of tipping their hand too early. Do not let this hold you back. If your plans are so easily thwarted that they fall apart just because your opponent knows about them, then they're not very good plans. Don't be afraid to build the organization you need and start contacting voters early. Be proactive and take the initiative when possible. Successful campaigns are built on bold action, not timidity.

> When able to attack, we must seem unable; when using our forces, we must seem inactive; when we are near, we must make the enemy believe we are far; when far, we must make him believe we are near. Hold out baits to entice the enemy. Feign disorder, and crush him. If he is secure at all points, be prepared for him. If he is in superior strength, evade him.

If your opponent knows exactly when and where you plan to send mailers, make calls, or knock on doors, they can maneuver to undermine you, by sending conflicting messages or flooding your target areas with their own campaign materials.

To avoid this, try to maintain a degree of unpredictability in your tactics and to keep your opponents guessing about your next move. Whether knocking on doors, attending meetings with PACs, or trying to earn endorsements, you must be careful not to reveal the intent behind your actions until your plans have succeeded. If your organization is strong, you must ensure that your enemies think you're weak and ineffective. If your opponents see weakness, they are likely to pounce. Keep your opponents guessing, and you'll maintain the upper hand.

Use bait as a weapon where possible. Lure your opponent into making the wrong move by pretending to be weaker, more disorganized, or less capable than you are. Then, when they have made too bold a move, take advantage of their lack of preparedness and seize the advantage for yourself.

A common example of appearing weaker is seen among larger campaigns, where the budget is not so tight. They hold on to large checks without cashing them and wait till the last minute to receive large donations from other PACs. The later you can wait to receive a donation, the later you must declare the donations to the state. Just don't wait too long and miss out on paying for something you need.

Don't challenge an opponent for the support of their base. Instead, try to win the votes they're not going after. If your opponent's organization is more robust or better trained than yours, you will need to compete elsewhere. You might consider reaching out to those who aren't likely to show up on their own or focus on registering new voters rather than relying on high propensity known independents. It's not worth your time to fight in their territory—go where they aren't. Beware that your opponent will also be trying to do this to you.

If your opponent is of choleric temper, seek to irritate him.

Definitely not as popular of a quote as "All warfare is based on deception.", but this is one of my favorite parts of the chapter. If your opponent is easily irritated, do your best to piss him off; engage such an enemy on social media and in the press.

A few years ago, one of my candidates faced off against one of those people. She was very active on social media and would publicly respond to anyone who engaged with her. By organizing a few local trolls, we kept her distracted, and we were able to coax out quite a few gaffes. The press was happy to pick up the story and gave lots of attention to the things she said about some of her fellow party members.

Pretend to be weak, that he may grow arrogant.

Faking unpreparedness like this can be a powerful tactic in its own right, but only if you are truly ready to take advantage of the situation. If you can lure your opponent into feeling overconfident and arrogant, it may give you the means to humiliate him or damage his credibility. But be careful, if you're not ready to take advantage of their cockiness, you might be the one who ends up humiliated.

If he is taking his ease, give him no rest.

When your opponent decides to take a break from the campaign trail, do not let him relax. As any experienced politico will tell you, campaign burnout is very real and very dangerous. Ensure it happens to your opponent and not to you.

Many years ago, one of my candidates impulsively took a family vacation abroad, unbeknownst to his campaign staff. He had enough of the campaign trail and was not coping particularly well.

While he was off enjoying sunsets and sandy beaches, his opponent seized the moment, launching an aggressive campaign in his absence. Press releases, public appearances, and targeted ad campaigns filled the gap my candidate left behind. By the time he returned, the damage was done. The public had seen his opponent tirelessly working while he was away. Do not let this happen to you.

If his forces are united, separate them. Attack him where he is unprepared, appear where you are not expected.

If your opponent is focused on just one aspect of the campaign, strike elsewhere. Shift your attention to the areas they're unprepared for and launch a surprise attack. This will force them to divide their resources and attention, leaving them vulnerable to your attacks. By targeting the voters and policies they don't expect, you can knock them off balance and take the advantage for yourself.

Now the general who wins a battle makes many calculations in his temple ere the battle is fought. The general who loses a battle makes but few calculations beforehand.

When it comes to planning, the devil is in the details. First, you need to set your goals and break them down into manageable benchmarks. Next, identify the steps that will take you toward those benchmarks. Break these down into individual actions that can be incorporated into your team's daily or weekly routines as recurring tasks. Put as much on your calendar as you can now. Consider everything that will go into reaching your goals.

Caitlin Huxley

Now, something that is incredibly important but that people do not like to hear: not every campaign is about winning on election day. In fact, with regularity, I consult with campaigns that are being run solely because nobody has challenged a particular incumbent in a few years or because they hope their race will cause the opposition to spend money defending a secure seat. This seems like a fine use of "deception", but only if you remember this when writing your plan and building your budget. These strategies can work, but they require careful planning and strict adherence to this strategy. Otherwise, the campaign can drift off course and end in disappointment.

I regularly see those candidates get caught up in the excitement. They fight to raise money, recruit volunteers, and otherwise run their campaigns as if victory includes being elected, violating the original spirit of their strategy. This sort of drift happens when campaigns do not have a clearly developed plan, or when candidates do not adhere to the plan they have written.

Similarly, candidates should be honest about their personal reasons for running, if only with their staff and close advisors. Building name recognition for future elections, expanding your business, championing an issue that you care about, and building up the local party are all valid reasons to run for office. But, if this is your goal and you veer too far away from it, and let your team drift toward the more alluring goal of winning on election day, you will either not achieve them at all, or you will severely draw your campaign off course in order to try to achieve them.

Finally, you should make sure to choose staff whose motivations align with your own. When I go to work for a candidate, I make it clear that my goal is to build up the party's strength for the long term. I want to work with candidates who want to win by identifying and training precinct captains, recruiting committeemen where they are missing, and working with the local party to grow a long-term structure. If I were to work for a candidate whose goals are different, then we would not see eye-to-eye and would be working at odds. Avoid this sort of mismatch.

II. Waging War

In the operations of war, where there are in the field a thousand swift chariots, and a hundred thousand mail-clad soldiers, the expenditure at home and at the front will reach the total of a thousand ounces of silver per day. Such is the cost of raising an army of 100,000 men.

"He who wishes to fight must first count the cost"—Ancient Chinese general Ts'ao Ts'ao is quoted as saying in a much earlier commentary on The Art of War. I believe this sums this chapter up very well. Campaigns are expensive, and they get more expensive with each passing year.

Each campaign's costs will differ and can vary wildly depending on the position you seek. Your expenses will grow with the size of your district, the quality of your staff, and the amount of time left.

When your organization is finally up and running near the end of the election cycle, you'll be paying staff salaries, buying meals for volunteers, giving out literature and signs, paying for ads, and a myriad of voter outreach methods. In fact, since most of your funds should be going to pay for direct voter outreach, and there is always more you could do, there's literally never enough money. If there is any money in your campaign's bank account, you'll spend it. Even the most grassroots of campaigns will be able to burn through any size war chest.

As I said, most of your budget should be spent directly on contacting voters. Because of that, it's easy for costs to inflate and for things that seem necessary to take priority over things that will contribute to your victory. This is especially dangerous for local campaigns or those with smaller budgets. Do not budget for so much collateral (yard signs, stickers, walk cards, etc.) that you cannot pay for the voter outreach they're used for.

This is exactly why a budget and cost comparison is so important to every campaign plan. Add up everything you'll need to pay for during your campaign and be sure to include buffers for miscellaneous purchases you may not have foreseen. This is your cost. The budget works backward from your expected total amount raised. Instead of what you expect to need to pay, focus on the amount of money you and your advisors believe you can raise.

Compare these two calculations and if they are far off, find a way to close the gap.

When you engage in actual fighting, if victory is long in coming, then men's weapons will grow dull, and their ardor will be damped. If you lay siege to a town, you will exhaust your strength. If the campaign is protracted, the resources of the State will not be equal to the strain.

Now, when your weapons are dulled, your ardor damped, your strength exhausted and your treasure spent, other chieftains will spring up to take advantage of your extremity.

Because campaign budgets are so hungry, you've basically got one shot at election. Campaigns fall into three main phases: org building, voter outreach, and GOTV (get-out-the-vote). During the first phase, costs will be minimal, and you'll likely raise more than you can spend. But, once you've started up your campaign machine, you can't stop it until the votes have been cast, and it will need an increased fuel supply the longer it runs. You will not have the time to lay siege to a district.

If you run out of cash, your organization will fall apart mid-campaign, and your staff will naturally leave you for other campaigns which can pay them. Sensing your weakness, donors will likewise leave for other campaigns. Your coalitions will be the last to leave you, but once the money dries up, so too will their support for your race.

This sort of thing can be seen most often in presidential primaries. The first candidates to drop out have their campaigns gobbled up by those who remain. But before that happens, you'll notice a drop in their volunteers' and donors' enthusiasm. The loser might endorse one of the remaining candidates and hope to be incorporated into their campaign, but as I'm sure you've seen, there is no guarantee they will give him any position of importance.

Campaigns have far more fair-weather fans than most would like to admit.

Thus, though we have heard of stupid haste in war, cleverness has never been seen associated with long delays. It is only one who is thoroughly acquainted with the evils of war that can thoroughly understand the profitable way of carrying it on.

The skillful soldier does not raise a second levy, nor are his supply wagons loaded more than twice. Bring war material with you from home, but forage on the enemy.

At the beginning of your campaign, you will likely start by self-funding. This is done either completely by yourself or with a group of donors who are very close to you. The largest and most reliable donors will usually expect to see a large amount of money in your campaign account before they will believe you are a serious candidate and before they will open their own wallets. It is not always possible to approach these same donors again, as there are often legal restrictions on the maximum amount of contributions they can give. Even without the law, donors will have budgeted only so much for political giving.

Relying on your supporters to fund a fully operational campaign for any longer than necessary will strain their financial resources and diminish their willingness to contribute further. You do not want to lose the goodwill you've built up.

Although this approach may not always be feasible, it is best to find new supporters from those who have been persuaded to switch allegiances from your opposition. If your coalition has successfully turned a large segment of voters away from your opponent, you should try to recruit heavily from this new base of supporters. Every dollar and volunteer earned this way is worth double because your opponent can no longer count on it. This is foraging on the enemy.

With all this talk about constantly expanding your reach to new supporter groups, candidates can't forget about the people they started with. Do not take them for granted. Do your best to activate them for donations, volunteer, improve your name ID, etc. But do not press too hard on them without bringing others in to help lighten the load.

> *Now in order to kill the enemy, our men must be roused to anger; that there may be advantage from defeating the enemy, they must have their rewards.*

> *This is called using the conquered foe to augment one's own strength.*

While in modern political campaigns, we are not killing each other, and thus field staff and volunteers do not perhaps require the same level of motivation as they would in actual warfare, they must still prefer your candidacy to your opponents. And it does work so much better if they've been roused to anger. In my experience, volunteers who start off hating my opponent are always the ones who are willing to work the hardest and the longest hours. In order for them to want to give up their free time and knock on doors in the rain or snow, they are going to need a pretty good reason. Be sure you're giving them one.

One of the target groups outlined in the famous fundraising manual "Making the Dough Rise" is the axe-bearers, described as those who already have a fiery hate for your opponent. The books "Start with Why", and "The Tipping Point" also go deep into covering this point when it comes to consumer and voter sentiment. A candidate with a passionate story of why they're fighting is significantly more successful at rousing people to action than a well-detailed 5-point plan will ever be.

Additionally, your staff must be able to see the rewards of their actions directly. As I write this, I hear in my head the almost mocking question of a candidate I advised a few years back; "you mean more than helping to get me elected?". Yes, John, obviously more than that.

Candidates tend to get a little hyper-focused on their race and forget that their supporters have many options in what to do with their free time and which campaigns they will volunteer on or donate to. You will need a lot more than your name on the ballot to bring them in and keep them engaged.

So, be sure to pay your staff well and ensure they feel they are working for a cause they are passionate about. If a staffer recruits a large number of volunteers or finds some max-level donors, pay them back for their success. Win bonuses (paying staff 1–3 months of additional salary upon victory) are very common within the industry for this exact reason. Compare your pay and benefits package to the other campaigns in your party, and see how you can differentiate yourself from them.

Similarly, in a primary election, you will be facing off against your own party, and when you win, you'll need to be able to adopt and absorb your opponents' infrastructure into your own organization. This is the only way to wield the full strength of your party's resources. If one of your opponents seems ready to drop out, it would be wise to try to win their endorsement and encourage the candidate to lend their supporter base, resources, and skills to help you defeat your mutual enemies on the other side of the aisle. This is using your conquered foe to augment your own strength.

In war, then, let your great object be victory, not lengthy campaigns.

As the saying goes, work smarter, not harder. Always remember that your campaign is a means to an end. So, for every action you take, consider how it will bring you closer to your end goal. This is the principle of conserving your economy of actions. If you can minimize the amount of time it will take to achieve your goal, you will be more likely to be able to afford your campaign.

To this end, in your first conversations with party leaders, campaign advisors, and campaign staff, bring up both the cost and the budget. Pay careful attention to what these experts think you must spend on and how much you will need to raise. Write down these numbers, average them up, and include a comfortable margin of error. This is the number you should seek to raise before the outreach phase of the campaign cycle, as you will want to be able to buy your campaign materials and pay for mailers early when they are cheaper. You will be glad you did when you can coast through your campaign without constantly stressing about the costs catching up with you.

On a related note, if you think there is not enough time left to win the election, then it is worth considering extending your plans. Take more time to prepare, raise funds, build your volunteer base, and maybe make a softer run this cycle aimed at org building. Then next election cycle, you will be equipped to run with all your force. You cannot run a 4-year campaign for a 2-year position, but you can take 2 years to prepare while pretending to run.

When using this tactic, just beware of becoming the chronic candidate; this repeat offender exists in every state and is a bane to their party. Some of them have been running for the same seat for the last 10 years, while others jump around, first running for Congress this year, city council the next, then trying for state senate. Make sure you run with purpose and do not expend resources unless you see a serious opportunity for victory.

In your election, you should let your great object be victory on election night and not lengthy campaigns.

Part 2: Strategy

"The enemy is only another algorithm. Simplify it, solve it, and set it in its place."

Avinash Dixit—The Art of Strategy

In political campaigns, as candidates adjust and refine their strategies, the race often reaches a Nash equilibrium—a term from game theory referring to a point where no change seems beneficial to either side. Knowing whether your campaign has reached this equilibrium serves as a check, guiding you on when to maintain your current approach and when to seek a new strategy to improve your chances.

In this part of the book, we will delve into the strategic aspects of campaigning, viewing the political landscape as a complex game where the moves of all players interact and shape the outcome. Viewing campaigns through the lens of game theory will allow you to gain some mathematical clarity into how you set goals, estimate your likelihood to succeed in reaching them, and decide which tactics best serve your overarching campaign plan.

Here, again, I will underscore that winning is not merely about victory in individual battles, but about achieving your broader campaign goals, which must be defined according to your personal set of circumstances and your desired outcomes. By applying the principles of game theory, you can pursue your desired outcomes, no matter what your unique goals may be.

III. Attack by Stratagem

* * *

Section 1: Tailoring your strategy to the circumstances

In the practical art of war, the best thing of all is to take the enemy's country whole and intact; to shatter and destroy it is not so good.

Hence to fight and conquer in all your battles is not supreme excellence; supreme excellence consists in breaking the enemy's resistance without fighting.

Clausewitz says in the opening chapter of "On War", his well-known thesis, *"If our opponent is to be made to comply with our will, we must place him in a situation which is more oppressive to him than the sacrifice which we demand."*

Imagine the hypothetical situation where your political opponents comply with your every will. You would have them drop their campaigns against you and instead donate their resources to help you win. As outlandish as this might seem, take a moment to consider that idea, and try to devise a plan to motivate them to act in the way you want. It may seem silly now, but depending on your opponent and the situation in your district, it may be more doable than you think.

If you could dream up the perfect campaign strategy, it would be to somehow win without all the expense and effort of a full-blown campaign. If you can somehow develop this perfect strategy, then you can supersede the need to develop a field program, run ads, or send mailers in the first place. Often enough, if you can come out of the gates strong and show yourself as the clear winner for an open seat, you may never even be challenged. This is often the case for many non-partisan races where parties are not disclosed to the average voter, and in extremely partisan districts where candidates from the minority party don't even bother to run and the primary is essentially the general.

At the other end of the spectrum, you should know that winning a single election is not going to be enough to secure your position forever. Some day you will be the hated incumbent to some group of voters in your district, and a challenger will come for you. You can stave this off by making victory seem impossible against you.

Just as was discussed in the previous chapter, with the hypothetical of trying to convince your primary opponents to join your cause after they've dropped out, you should also try to find a way to incorporate your general election opponent's supporters, donors, and other resources into your own if you can. This is much more difficult, but ultimately much more important, than with your primary opponents. This is because one vote taken from the opposing party counts double, and because if you do not, they will eventually run again or lend their support to your opponent in the next election. If you do not want to run a full campaign every election cycle, then you will have to employ a clever stratagem.

> Thus the highest form of generalship is to balk the enemy's plans; the next best is to prevent the junction of the enemy's forces; the next in order is to attack the enemy's army in the field; and the worst policy of all is to besiege walled cities.

> Therefore the skillful leader subdues the enemy's troops without any fighting; he captures their cities without laying siege to them; he overthrows their kingdom without lengthy operations in the field.

> This is the method of attacking by stratagem.

If you are an incumbent, it is only a matter of time before you are challenged. When you first ran for office, you became the chosen candidate from your party, and won the affection of the voters; then you ran the campaign needed to win the election.

Now, in the same way, your opponents will scheme against you. Regardless of if you're an incumbent or a challenger, you should do what you can to anticipate their plans and move to mitigate them before they can get off the ground. Chapter 13: "The Use of Spies", includes more about discovering their plans, but for now, just imagine yourself in their shoes. What would you be doing if you were them, and what is the best defense against that? A major part of success in this activity is being honest about your weaknesses and your opponent's strengths. This is why I suggest you start by completing a SWOT analysis and filling in a Leesburg Grid from your opponent's perspective.

If you can't stop other people from entering the race, you should instead try to prevent them from building their coalitions, recruiting their volunteers, and soliciting their donors. Think about who their potential allies are and what you can do to isolate them. Continuing the example of the Circles of Benefit from Emily's List's "Making the Dough Rise", make a list of the groups supporting your opponent and come up with a few ways each might be pacified. The candidate's personal circle is unlikely to be dissuaded from donating or volunteering. Still, potential supporters from their ideological, axe-to-grind (our direct opposition), and power circles are increasingly easy to mollify. The further divided their supporters are, the more places where wedges may be driven, the better your chances.

Just as I've advised you not to take your base for granted, you should watch your opponents for signs that they have ignored theirs. If they are not in communication with part of their core base of supporters, that voter group may make a good target for your efforts. If they're not being asked for money, to volunteer, or to be a closer part of your opponent's campaign, perhaps they would like to be a part of yours instead.

Because they have had years to solidify themself, the worst of the possible scenarios is to have to face off against an incumbent. This is Sun Tzu's Walled City. Incumbents have serious advantages which are very difficult to overcome. They have resources you cannot steal, experienced staff who've won a victory in-district, and they have strong ties to the community, solidified by years of trading favors, sponsoring events, and championing local causes. Thus, you should not try to challenge incumbents unless something unexpected has happened and their influence or stability in the community has waned. Perhaps redistricting has changed the partisan makeup of their district, or you've received reliable intelligence indicating the seat is weak.

If you take a step back and look at the political landscape in your state, you will surely see plenty of impatient candidates running against well-entrenched incumbents who are nearly impossible to remove. Just like with any normal campaign, they raise money and recruit volunteers from supporters within their party, knock on doors, send mailers, run ads, and make speeches at fundraisers. But because they were never in a position to win, they spent a lot of time, energy, and money with nothing to show. The effects are disastrous for their party and the supporters who believed in them. And to top it off, most of the time in situations like this, the candidate is never seen nor heard from again.

To avoid this pitfall, you can plan carefully, try to foil your opposition's plans before they can build their campaigns against you, and refuse to go on the offensive except when you have a serious advantage. You should never run campaigns you cannot win. Though, as stated before, I definitely believe that "victory" should be more widely defined than just "being elected." For example, Libertarian and Green party candidates might consider it a win to receive 3–5% of the vote statewide, or the champion selected by a single-issue organization might be happy to step down once the Governor signs his bill into law. You must decide your own victory conditions; just be sure they are achievable.

If you're to be a skilled strategist, you'll need to devise plans to win elections without running costly races against impossible odds. Pick your battles wisely. In an ideal world, you can win your race without running at all. This is what Sun Tzu calls winning by stratagem.

* * *

Section 2: Estimating your chances for achieving victory

If your race is tight, then to be elected you will need to compete with your opponents in the field. If you cannot risk letting go of a single advantage, then you should know that your performance will depend on the size of your organization. Suppose your voter outreach capabilities are equal to your opponents. In that case, your victory will hinge on your success in engaging with voters at the doors, on phones, in social media, via texting, or whatever technology brings next.

Over the last few election cycles, I have heard more and more dissenting voices talking about the shrinking impact of traditional campaign activities on election day outcomes. With the advent of every new technology made for contacting voters, the importance of volunteer-based voter outreach is said to be lessened. Whether or not this is true to some degree, volunteers are still a resource available to you, and ignoring them in favor of the shiny new thing is absolute folly. If you don't use your volunteer base on voter outreach, you're purposefully giving up that advantage to your opponent. I once saw an election end in an actual tie, with both candidates receiving the exact same number of votes; the state literally flipped a coin to decide the race. A ridiculous outcome! Obviously, the loser of that coin toss wishes he'd have spent just one more hour knocking on doors, and the winner wishes he hadn't left it to chance.

Just as with the money in your budget, you should be thinking about how you can use your human resources most efficiently. Do your calculations, and figure out how many voters you will need to identify, convince, and GOTV. Then figure out how long it will take for your team to do that much work. Compare those numbers and look at whether you have a deficit or abundance. Adjust your campaign plans accordingly. This might seem a bit overwhelming, but there are plenty of resources available. My site, HuxleyStrategies.com is filled with free guides advising on best practices.

It is the rule in war…
> *…if our forces are ten to the enemy's one, to surround him*
> *…if five to one, to attack him*
> *…if twice as numerous, to divide our army into two*
> *…if equally matched, we can offer battle*
> *…if slightly inferior in numbers, we can avoid the enemy*
> *…if quite unequal in every way, we can flee from him.*

Hence, though an obstinate fight may be made by a small force, in the end, it must be captured by the larger force.

If after running your calculations, you believe your base of voters vastly outnumbers your opponents' base and your team can do the required work 10x over, then it probably doesn't matter what you do or who you spend your time talking to. In fact, if you know you can't lose based on the numbers, then you might as well spend your time talking to voters from your opponent's party to peel off the outliers, identify those who would vote for you, and build your cross-party base of support to limit potential opponents in future elections. If you're in this position, it is usually because your party controls your district. Either you've been hand-picked by party leaders, or you've just won a primary election, and if your general election opponent exists, he only has token support from his party.

If you have a sizable lead, but nothing is yet set in stone, you can probably still afford to split your attention a bit. Suppose that spending half (or some other percentage) of your resources in the field is enough to achieve victory. In that case, the remaining percentage is free to focus on your other long-term aspirations and can be sent off to try to achieve secondary goals. Think about what else you might want to accomplish, but give preference to those that will still contribute to election day victory while keeping a healthy margin of safety. The last thing you want to do is spread yourself too thin and fumble your chances to win.

If you believe your team is just about equal to the task ahead, then you should focus your efforts entirely on outworking your opponent. There is no leeway for distractions here, so wherever possible, only take the most economical actions. Outreach only to the voters you've targeted as being the most persuadable, and GOTV only to those known supporters who will not show up at the polls unless you drag them there.

If your expected vote deficit is only slightly smaller than your ability to overcome it, you should *avoid the enemy*, as Sun Tzu says. In a perfect world, this might mean waiting for a more favorable year or building up your organization before you run. However, this is not always possible, and if, for whatever reason, you MUST run this election, then you should only try to target the voters your opponent is not. If you cannot hope to compete to win their votes, don't; compete for other votes instead.

If, after honest reflection on your chances, you feel your organization is incapable of overcoming the obstacles, then you will need to avoid engaging with your opponent or his supporters in any way. A smaller organization can put up a fight but will eventually lose on election day. You might be able to get what you want from running, but only if what you want doesn't involve being elected. In this situation, I would try to refocus my efforts on a more creative interpretation of "victory."

Remember that how you define victory is up to you, and what goals you hope to accomplish. If your current idea of victory is impossible, then it is obvious that you must change your goals. But, if your goals are too easy, then it is also probably wise to adopt new ones. If you have resources in abundance, think what else you could accomplish; what ally you can aid, what higher office you can set your sights on, what bill you can help turn into law. Just don't overestimate your chances and accidentally stretch your lines too thin.

* * *

Section 3: Avoiding Common Mistakes

There are three ways in which a ruler can bring misfortune upon his army:

    (1) By commanding the army to advance or to retreat, being ignorant of the fact that it cannot obey. This is called hobbling the army.

    (2) By attempting to govern an army in the same way as he administers a kingdom, being ignorant of the conditions which obtain in an army. This causes restlessness in the soldier's minds.

    (3) By employing the officers of his army without discrimination, through ignorance of the military principle of adaptation to circumstances. This shakes the confidence of the soldiers.

Like the other lists in this book, there are obviously far more than only three ways to invite disaster into your campaign, but Sun Tzu chose these three for good reason; they are the big ones. If you stick around in politics long enough, you'll see these things happen repeatedly. Avoid making these mistakes yourself, and you'll avoid the most common sources of catastrophic failure.

Firstly, don't give impossible commands. This might seem obvious, but you'd be surprised by how often this sin is committed.

When races are close, and candidates are scrambling for every last vote, it is not unusual to see campaigns asking their staff to hit impossibly high numbers or work impossibly long hours. Likewise, when election day is very near, yet the volunteers have not been adequately trained or are not equipped with the proper tools, inexperienced managers will often send them out unprepared anyway because time is short. As you can imagine, this does not lead to success.

Secondly, do not try to run a campaign like you would run a business. Your staffers are unlike typical employees, and campaign offices cannot maintain regular nine-to-five type office hours. Early in the campaign cycle, your staff will essentially be part-time, and when the election draws near, they'll be working overtime. Early on, staffers have nearly unlimited vacation days, as long as work gets done, but there are no paid sick days, no maternity leave, 401k, and no benefits package. Most professional staff are paid as consultants or individual contractors, while interns and volunteers are often not compensated at all. Treating any of your team like an employee is certain to cause agitation, or at least give them the wrong idea about what is expected of them

Third and finally, be sure to use the right person in the right place. Experienced staffers will develop a specialty over time. Even in smaller campaigns, where many of the positions will be rolled together, it is still never an effective strategy to assign staff duties outside their wheelhouse. If you hire a campaign manager with a background in field and data, you'll get much better return on your investment having them write your campaign plan than if you asked them to do fundraising. Nobody can be good at everything, so you will need more than a single person working on your campaign, no matter how small it is. Remember that a jack of all trades is a master of none, even if they're often better than a master of one.

> *Thus we may know that there are five essentials for victory:*
> *(1) He will win who knows when and when not to fight.*
> *(2) He will win who knows how to handle both superior and inferior forces.*
> *(3) He will win whose army is animated by the same spirit throughout all its ranks.*
> *(4) He will win who prepared himself, waits to take the enemy unprepared.*
> *(5) He will win who has military capacity and is not interfered with by the sovereign.*

If you run your campaign inefficiently or ineffectively, then even if you've developed the best plan in the world, you may still find a way to throw it all away. Keep to Sun Tzu's five essentials, and you'll be able to avoid the most common pitfalls:

Do not run if you are not likely to be victorious. Be sure you know whether you have an advantage or disadvantage in terms of numbers, and act accordingly. Ensure your team's morale is high and that they remember why this race is worth fighting for in the first place. This goes hand in hand with the essential #1. Remember that warfare is based on deception and use this to your advantage whenever possible. Point #5 is directed at candidates. Do not interfere with your campaign staff. Presumably, you hired them because they know what they are doing, so you should trust them to do what they need.

If you know the enemy and know yourself, you need not fear the result of a hundred battles.

If you know yourself but not the enemy, for every victory gained you will also suffer a defeat.

If you know neither the enemy nor yourself, you will succumb in every battle.

If you have done your research so that you know everything there is to know about your district and opponents, and if you are honest with yourself about the inherent strengths and weaknesses of your campaign, then you will be able to choose which elections to run for. As a result, you will have a solid chance to win every time. This is what it means to know your enemy and know yourself.

The perfect strategy is tailored to the specifics of your district, expertly maneuvers past your potential opponents, and secures you against all future attacks. But it is also realistic. It takes into account your actual chances of winning and doesn't try to do the impossible. Instead, it makes the best use of the resources you have available and employs efficient actions wherever possible.

When employing your strategy, don't ask too much from your team, and learn from previous campaigns to avoid their mistakes.

IV. Tactical Dispositions

Tactics are the practical steps, the actions and maneuvers, that bring your overarching strategy to life. They represent the mechanics of your campaign, the day-to-day activities, the actions taken in the field that keep your operation moving. Unlike strategy, which paints a broad picture of your path to victory, tactics revolve around immediate actions such as courting specific donors, securing a position within local community organizations, or how you will respond to your opponents' latest moves. By understanding and implementing both defensive and offensive tactics effectively, you set the groundwork for your campaign's success.

In essence, strategy lays out 'what' your campaign aims to achieve and 'why,' while tactics explain 'how.'

* * *

Section 1: Your strategy defines your tactics

The good fighters of old first put themselves beyond the possibility of defeat, and then waited for an opportunity of defeating the enemy. To secure ourselves against defeat lies in our own hands, but the opportunity of defeating the enemy is provided by the enemy himself. Hence the saying: One may know how to conquer without being able to do it.

As I've repeated a handful of times now, you shouldn't run if you can't win (of course, allowing for your own individual victory conditions). This is the easiest way to put yourself beyond the possibility of defeat. Doing your research and deciding to run only when you have everything needed to secure a victory is how you live up to the second part—for an opportunity of defeating the enemy. As a challenger, this is an easy decision to make, just waiting around for a scandal or change of demographics so that you have an opportunity to strike. But what if you're the one in the hot seat?

As an incumbent, maintaining the support of your voter base is up to you. Build good relationships with your donors and other stakeholders, secure your position within community organizations, and ensure your reputation is above reproach. However, your challengers will not show themselves until they think it is time, and the opportunity to drive a wedge between them and their supporters will have to come from their own words or actions. Watch the pretenders to your seat carefully. Discover who among them seems to be positioning himself against you. Remember that supreme excellence consists in breaking the enemy's resistance without fighting. So if given the opportunity, you should take steps to stop them from challenging you.

How much easier it becomes to watch your opponent once he has revealed himself. If you do end up with a challenger running for your seat, use spies to monitor him, as outlined later in Chapter 13. Does an opposing candidate neglect their walk program? Did he do or say something to offend some voters? It is your job to find the hole in his otherwise perfect strategy and exploit it.

The difficulty is in positioning yourself to be able to exploit his weakness. For example, if you want to try driving a wedge between him and one of his supporter groups, you would have to have already established credible relationships within that community yourself. To steal a donor after he offends them, you'd have to already be in communication with them. A healthy amount of this preemptive effort can be a prudent tactic, just be careful not to become obsessed. If you spend all your time watching what your opponents are doing, you'll find yourself forgetting to build your own campaign, to court your own voter base, and to fundraise from your own donors. More on Positioning can be found in Part 5.

Like Sun Tzu writes, *"One may know how to conquer without being able to do it"*. You may see a clear path to victory, and not be able to follow it, as long as your opponent keeps it well guarded. You will have to wait for him to let his guard down, or find a way to force him to focus his attention elsewhere.

Security against defeat implies defensive tactics; ability to defeat the enemy means taking the offensive. Standing on the defensive indicates insufficient strength; attacking, a superabundance of strength.

A big part of a successful campaigning strategy is choosing your battles. So prepare your own defense first, then only attack when an opportunity presents itself and you're capable of taking advantage of it. I have observed a growing trend of candidates who are convinced that in the fast-paced world of modern politics it is more important to stay open and respond to circumstances as they arise than to bother with a plan that will end up changing anyway. This causes them to try to remain too open to possibilities, forgetting to first *put themselves beyond the possibility of defeat*. They are missing that crucial first step. These candidates end up with too few volunteers, low donor excitement, and overall not enough fuel to keep their machine running until election day. This point is greatly expanded in the following chapter on Energy, but here let's suffice it to say that you should not go on the attack until you are strong. Time your tactics appropriately.

Making no mistakes is what establishes the certainty of victory, for it means conquering an enemy that is already defeated. Hence the skillful fighter puts himself into a position which makes defeat impossible, and does not miss the moment for defeating the enemy.

Thus it is that in war the victorious strategist only seeks battle after the victory has been won, whereas he who is destined to defeat first fights and afterwards looks for victory.

Once you've discovered the optimum strategy for your district, written out your campaign plan, and built your organizational resources up to the point that you can accomplish your goals, you must still be careful to make no mistakes. Just as you are watching your opponent and waiting for him to provide you an opportunity to attack, he will likewise be watching you. If you make a gaffe, mistarget your phone banks, fail to court the right influencers, or otherwise deviate from perfection in any way, you are giving your opponent a free advantage over you. If it's severe enough, you may end up giving away the entire election. Every course of action you consider is a possible tactic, and should be evaluated for its impact on your strategy and the success of your campaign. If you act impulsively, out of pride, or without considering the consequences you could end up making a mistake and you may accidentally torpedo your own chances. Nothing can save a good campaign from a bad candidate. Think then act, not the other way around.

* * *

Section 2: Evaluating Tactics and Maximizing Chances

The consummate leader cultivates the moral law, and strictly adheres to method and discipline; thus it is in his power to control success. In respect of military method, we have, Measurement, Estimation of quantity, Calculation, Balancing of chances, and Victory.

Measurement owes its existence to Earth; Estimation of quantity to Measurement; Calculation to Estimation of quantity; Balancing of chances to Calculation; and Victory to Balancing of chances.

Remember from Chapter 1 that the *Moral Law* is the overlap between the issues you highlight in your campaign, and the issues the voters of your district care about. Advantage here will help you recruit volunteers, secure donations, and win votes from key demographics. *Method and Discipline* is the training and organization of your staff and volunteers, and is the defining factor in whether your team is a ragtag team of volunteers or a professionally managed staff.

I interpret *Measurement*, *Estimation*, and *Calculation* as the necessary research with which you should begin each campaign, and the iterative process of developing your campaign strategy based on that research. This begins with *Earth*, which are the physical dimensions of your district and its break-down into counties, townships, municipalities, and precincts. It is also the issues up for debate in the current political climate, from which you might derive an advantage or on which you might be attacked.

Balancing chances is about being honest with yourself and your staff about your likelihood to succeed with any given tactic and how it might affect your overall strategy. Remember here to calculate not only your own chances, but also your opponents. Think of how they might move to disrupt your plans, and how you might change tactics if they try.

Now, many candidates approach their campaign with assumptions about their district, which issues voters care about, and how to win. Doing so causes them to build off of those assumptions as if they were facts, and thus end up with a plan not based in reality. In this way, you can hopefully see how this mistake can cause catastrophic failure down the line. Do you best to avoid working off assumptions

> *A victorious army opposed to a routed one, is as a pound's weight placed in the scale against a single grain. The onrush of a conquering force is like the bursting of pent-up waters into a chasm a thousand fathoms deep.*

Sun Tzu tells us that if we can set up a solid defense, and stay open for the right time to attack, we will be successful. Bide your time building your base of supporters, supporting other candidates and causes you believe in, and then declare your candidacy only when an office is ripe for the taking. Your tactics will succeed easily, if you only choose those which are easy to succeed in.

Go after the low hanging fruit. Pick up as many easy advantages as you can find, and leave only the most difficult path available to your opponent. In your campaign, you should be looking for the tactics with the highest return on your investment of time and effort. The Pareto Principle, developed by Joseph Juran, and based on the work by Italian economist Vilfredo Paredo, sums this up well: *80% of all results come from only approximately 20% of actions*. You may have heard this referred to as the 80/20 rule.

By focusing on the 20% most influential actions, a campaign can allocate resources more efficiently, avoiding wasted time and energy on initiatives that offer limited returns. This refined focus can be the difference between a campaign that spins its wheels trying to be effective and one that is both effective and streamlined.

Part 3: Management Principles

"The best executive is the one who has sense enough to pick good people to do what he wants done, and self-restraint to keep from meddling with them while they do it."

Theodore Roosevelt

Roosevelt here lays the foundation for Part 3: "Management Principles," where we'll explore the the crucial part played by your campaign team. Elections are not just about messaging, strategic maneuvering, or statistical analysis. At their core, campaigns revolve around people—the staff and volunteers who give their time and energy to the cause, the voters you try to convince at the door and on the phone, and the groups of supporters that make up your coalitions.

To win, you will need to assemble a competent team that can execute your plans, and exhibit enough restraint to allow them to do their work without unnecessary meddling. We'll discuss the balance that leaders must strike between providing guidance and allowing independence, between maintaining control and fostering initiative.

Part 3 of this book will empower you with the management skills required to steer your campaign ship, emphasizing that effective leadership and management are invaluable to your collective success.

Here, we will explore the process of building your team strategically, placing the right person in the right position, so you can reach your goal. Next we'll delve into using your team effectively, emphasizing the need for efficient delegation and the importance of trust. Effective leadership isn't just about assigning tasks; it's about knowing your team members' strengths and weaknesses, assigning roles that play to their strengths, and providing support to help them overcome their weaknesses.

V. Energy

Mastering energy in a political campaign is about understanding the difference between potential energy (the energy stored in an object at rest) and kinetic energy (the energy of that object while in motion). In this chapter, Sun Tzu likens this to a crossbow string being pulled back, and to the bolt being fired. In your campaign it is the strength of the organization you build, and the way in which you use it once it's built.

As discussed in previous chapters, in your campaign, you should only be expending your energy at the right time toward the right purpose. If your goals are impossible (or improbable) to accomplish, if your actions are poorly timed, or if the effort you and your staff put behind their actions is less than it ought to be, then you will meet with certain failure.

Oftentimes new candidates and staffers start off with a rush of activity, but this is like sprinting at the beginning of a marathon; If you're not careful, you'll quickly wear yourself out, your staff and volunteers will lose their drive and some will eventually stop showing up. This in turn will cause you to put more pressure on your remaining staff, which will only perpetuate the cycle.

Burnout is much higher on campaigns than it is in the private sector. This is especially likely to affect new staffers, as they are not yet used to the rigors of political campaigning. Each election cycle I see campaigns run out of steam before election day.

As discussed in the previous chapter, victory comes from building up your position to be able to strike and then waiting for the opportunity of defeating the enemy to present itself. Mastering Energy is crucial to this task.

* * *

Section 1: Building your Team

Now the general is the bulwark of the State; if the bulwark is complete at all points; the State will be strong; if the bulwark is defective, the State will be weak.

Your campaign manager is the backbone of your campaign. This one individual will be responsible for nearly every aspect of the campaign, even more so than the candidate at times. If you are a candidate, hire carefully; choose the applicant who is most skilled and most likely to handle the pressure without cracking.

On a campaign, the candidate's network should become campaign property. The candidate cannot be the only one with access to their list of friends and family who have agreed to help out. Everyone with a financial or emotional investment in the campaign has a part to play. A good commander is capable of managing the entire team, so choose wisely.

To see victory only when it is within the ken of the common herd is not the acme of excellence.

What the ancients called a clever fighter is one who not only wins, but excels in winning with ease. Hence his victories bring him neither reputation for wisdom nor credit for courage.

Every so often you run across a campaign manager or consultant who boasts about their win number, or win ratio. This is a very deceptive metric, because it imparts no actual information. Given just that number, how should you evaluate this person's skills? In the campaign world people have become focused on winning to the point that it has become a detriment of the rest of the process. It is more important how a candidate wins than simply if they win. When hiring a campaign manager or a general consultant you should ask about the unique situations that existed in their past races, and how they positioned themselves to take advantage of the opportunities those situations presented.

If in training soldiers, commands are habitually enforced, the army will be well-disciplined; if not, its discipline will be bad.

Of course, no matter how much effort you put into hiring the right person, the wrong people will sometimes find their way in. Cut loose your unruly volunteers who argue with voters, and find ways to motivate your underperforming staff. If you can't motivate them, let them go. The higher up the offender is in your organization, the more important this lesson is. Do not fall into the trap of the Sunk Cost Fallacy; it is never too late to fix a mistake.

Regard your soldiers as your children, and they will follow you into the deepest valleys; look upon them as your own beloved sons, and they will stand by you even unto death.

If, however, you are indulgent, but unable to make your authority felt; kind-hearted, but unable to enforce your commands; and incapable, moreover, of quelling disorder: then your soldiers must be likened to spoilt children; they are useless for any practical purpose.

If you treat your staff and volunteers well, understand what makes them want to volunteer, and give them jobs they want to do, they will keep coming back. When I join a campaign, I tell every staff member under me that I want to help shepherd their careers. I do this to earn their loyalty, and because I see it as my duty to encourage their growth. I want to ensure that my staff are learning, and growing into their new role each cycle.

If you do not train them, or don't make sure they are given useful jobs, they will not be worth the time and effort that went into recruiting them. Too often, staffers are driven harder and harder each week until election day. They are subjected to early morning conference calls meant to shame those field managers who haven't hit their numbers. If you foster this sort of attitude toward your team, you may get the most of your staff as long as you continue to put pressure on them, but let up for a moment, and they will slack off. Then at the end of the campaign you will lose them to the private sector.

The control of a large force is the same principle as the control of a few men: it is merely a question of dividing up their numbers.

Just like an army subdivides itself into a hierarchical structure, giving ranks to each of its members, and relies on clear lines to keep troops supplied, trained, and organized, so too should you strive to organize your campaign team. Within each department, you should break down your staff by rank, with more experienced staffers in charge of newer recruits. Talent and experience are ten times more valuable than seniority. For this reason, I am always wary of promoting staffers who have held the same position for several campaign cycles. Lack of growth is not a good sign in a director.

In an election, the candidate puts his or her personal reputation on the line, and must entrust many people to acting as his surrogates. This often leaves candidates feeling tempted to personally handle every aspect of their race. But a candidate or campaign manager who tries to take on too much, or who micromanages his staff is going to have his hands full. Besides being a violation of *Method* and *Discipline*, campaigns are volatile and things are going to slip through the cracks.

With proper delegation, managing a statewide campaign follows much the same principles as managing a smaller race with only a handful of volunteers and one or two staffers. You give orders to your campaign manager and trust they are being carried out properly. By dividing your campaign into the proper departments and putting someone you trust in place to manage each, you can ensure there are no cracks for things to slip through.

The traditional way to manage a large statewide campaign is by dividing it along department lines and appointing in each department a "director" level manager. The common departments are field, coalitions (often called political), communications, digital, and finance. Further divisions are often found in the field department with each region being spun off into its own pseudo-autonomous organization, with their own regional director to oversee the various office managers who in turn manage the local volunteer coordinators. You should separate your field department from your communications staff, so that each arm of your campaign can work to achieve its specific goals. Deviate from this if necessary, but only if you have the experience to make sure you're doing so for strategic gain.

Fighting with a large army under your command is nowise different from fighting with a small one: it is merely a question of instituting signs and signals.

One of the easiest ways to motivate and drive your entire team is to keep in constant communication. This way, no matter how far they are from your headquarters, the folks in the field are able to support your central campaign plan as it changes in response to changes in circumstance.

In Chapter 7, Sun Szu talks about instituting systems of *Gongs and drums, banners and flags*. Obviously, nowadays we can do a little better than that, and our communication methods are slightly more advanced, but the purpose remains the same. Successful modern campaigns often use an online platform such as Whatsapp, Groupme, or Slack in conjunction with project management software such as Trello, Asana, or Basecamp. And of course, this is going to change as technology advances.

Just like with voter outreach, your volunteer contact methods should consider how your volunteers and supporters like to be contacted. Many younger folks don't answer their personal cell phone if they don't recognize the phone number but are likely to read any text they get sent. Others are the exact opposite. Know your audience, and choose your outreach method very carefully. Whichever tools you decide to use, they need to be capable of instantaneous communication and should be robust enough to allow for an entire organized campaign to pass along updates.

Just because this allows you to be in constant contact with your team, doesn't mean you always need to be. You should beware that too many texts, emails, or calls can lead to information overload, and could result in the important messages getting lost in the mix. You should restrict who has access to communicate with your entire team, and make a plan of what can and should be discussed. Proper use of the channels of communication preferred by your team will help you direct your energy where it is needed, when it is needed.

If our soldiers are not overburdened with money, it is not because they have a distaste for riches; if their lives are not unduly long, it is not because they are disinclined to longevity.

On the day they are ordered out to battle, your soldiers may weep. But let them once be brought to bay, and they will display the courage of a Chu or a Kuei.

Remember that your staff chose to work on campaigns for a reason. If they were motivated only by money, they would be working in the private sector. They are not here because they dislike money, and they don't work tirelessly for long hours, forgoing sleep in the weeks leading up to the election because they are natural insomniacs who get bored with too much down time.

I know that I am not the only staffer who has literally cried from the high stress of a campaign job. Similarly, I cannot be the only one who fantasized about having a normal nine-to-five job, or one that pays time and a half for overtime! Your team is here because they believe in you and are devoted to the cause. Manage and motivate your team properly, and they'll show you why they chose political work.

* * *

Section 2: Managing your Team

To ensure that your whole host may withstand the brunt of the enemy's attack and remain unshaken—this is effected by maneuvers direct and indirect. That the impact of your army may be like a grindstone dashed against an egg—this is affected by the science of weak points and strong. In all fighting, the direct method may be used for joining battle, but indirect methods will be needed in order to secure victory.

This is one of the more interesting, but difficult to grasp, parts of Sun Tzu's manual.

Let's say your opponent makes a gaffe. Your immediate reaction might be to strike while the iron is hot. Your first thought may be that you would like to expose this mistake to the world. Perhaps strategically it would even be good for everyone to hear about it. The most direct maneuver here would be to post a tirade on facebook, send out a mailing to the district, or run a commercial highlighting this mistake. Indirect maneuvers would be having an allied organization do this instead. Not only will the impact be greater because it comes from a member of the community, but you'll be able to keep your hands clean. (Attacking in this way is discussed more in Chapter 12: "Attack by Fire.") You may be able to attack your opponent in the court of public opinion, but to win your race, you will need to include more sophisticated tactics into your strategy. Consider this when building your organization.

There is only one way to reliably win votes, and it boils down to guessing who is likely to be persuaded by your message, approaching them in the way they are most comfortable being approached, and making sure they actually vote. Will your campaign focus on approaching the masses by attending festivals and walking in parades? Or will you and your team directly contact voters you've segmented out? Will you focus only on persuading high propensity voters, turning out low propensity supporters, or will you try to register new voters altogether? By using the right mix of tactics, at the right times, you can concentrate your efforts where your opponent is not ready to defend.

To be successful at this requires that you do your homework on the district, the demographics, and historical vote trends, and that you know yourself and your opponent. Winning also involves a decent amount of luck, but by focusing your energy on achievable goals, timing your expenditures to when they'll have the greatest impact, and employing your troops in an organized and efficient manner, you can certainly tip the scales in your favor.

Consider also that you cannot do all of this alone. No matter the tactics you decide on, you will need staff, volunteers, and a coalition large and capable enough to pull it off. Once you know what tactics your strategy will require, recruit and use the right men for the job.

The onset of troops is like the rush of a torrent which will even roll stones along in its course. The quality of decision is like the well-timed swoop of a falcon which enables it to strike and destroy its victim. Therefore the good fighter will be terrible in his onset, and prompt in his decision.

Energy may be likened to the bending of a crossbow; decision, to the releasing of a trigger.

If building and training your organization properly is where half the battle is won, then the other half is won with proper timing and self restraint. Just as you will be trying to make the most of your opponents gaffes, they will be twisting your words to make it seem like it's you who's blundered. They'll target your supporters with inflammatory messages, and otherwise try to goad you into action before it's time. Self control is the only way to keep yourself from acting before the time is right.

Every election cycle, we see headstrong candidates ignore the advice of their staff, or worse, be prodded into action by them, and start running ads or sending mail too early. These candidates inevitably run out of money long before the bulk of voters have cast their ballots. Just as you are holding out bait for your opponent, you should not take theirs. Do not be pressed into action before it is time. Being early is almost as bad as being too late.

Amid the turmoil and tumult of battle, there may be seeming disorder and yet no real disorder at all. Simulated disorder postulates perfect discipline, simulated fear postulates courage; simulated weakness postulates strength.

Thus one who is skillful at keeping the enemy on the move maintains deceitful appearances, according to which the enemy will act. He sacrifices something, that the enemy may snatch at it. By holding out baits, he keeps him on the march; then with a body of picked men he lies in wait for him.

Remember that *"all warfare is based on deception"*. Through direct and indirect methods, you'll try to lull your opponent into a false sense of security, with him ideally believing your campaign is disorganized and ineffective. However, to pull this off effectively, your team must actually be well disciplined and in perfect marching order. If you want your opponent to attack you on an issue during a debate, you're going to have to play it like you don't want him to bring up the subject. You're going to have to work at least twice as hard in preparation if you want to play him for a fool. This will take your whole team, working together

When trying to pull off a feinting maneuver like this, don't be too hasty, and don't overdo it. I'm reminded of the many times I've heard clients say that we should not begin contacting voters yet because we don't want to tip off our opponents. This is ridiculously counterproductive! Instead of keeping your team out of the field to keep your strength a secret, you should seek to use your strength, but focus on what you can hide from view. It's hard to know how much effort you've put into volunteer recruitment and training, or how professional and disciplined your staff is.

Only when you are satisfied that your metaphorical crossbow is loaded and aimed perfectly, should you pull the trigger.

> *The clever combatant looks to the effect of combined energy, and does not require too much from individuals. Hence his ability to pick out the right men and utilize combined energy. When he utilizes combined energy, his fighting men become as it were like unto rolling logs or stones.*

When you are searching for volunteers or donors, make sure to go beyond their checkbooks or their ability to knock doors. If someone on your list is a graphic designer, an accountant, or even just makes a great batch of cookies, then they have added value to your team. Think about how each person's unique talents can be leveraged to make the largest impact.

Rely first on the strength of your organization as a whole, and only then consider the individual talent that might maximize your chances. Do not drive the individuals on your team too hard, and do not put them in situations where they are not strong. On smaller campaigns, you often see all efforts shouldered by a single campaign manager, or a few regular volunteers. This is too much and too important to put all on one person. Real life might get in the way, and out of nowhere you may lose one person, and with them, you're whole campaign. If you are structuring your campaign around the individual instead of around your campaign as a whole, you're courting disaster.

Prohibit the taking of omens, and do away with superstitious doubts. Then, until death itself comes, no calamity need be feared.

If I had a dollar for every slanted poll I've seen a candidate take to heart, or for every time a volunteer declared that this year would be a wave year for our party, I'd be able to retire today. If your team does not believe that their hard work is the only thing keeping defeat at bay, they will not be inclined to work hard for you. As election day nears, the number of lies we tell ourselves about our chances increases exponentially. Do not allow your team to believe that victory is assured.

As a campaign stretches on, it is common to fall into the trap of too much positive thinking. Internal polls will always skew your way, and volunteers are always so optimistic. I often find myself saying things like "We have a real shot at this" to spur my team on, and after a while, I believe my own white lies. So be honest, at least with yourself, about the factors at play in your election.

If the calculations show that the opponent holds all the cards, it may be wise to consider running for a different position instead. Don't let optimism blind you to the reality of your situation.

* * *

According to military strategist, B.H Liddel Hart, there is a Maxim put forth by Napoleon which I think describes the campaign cycle quite well. He says that *"The whole art of war consists in a well ordered and prudent defensive, followed by a bold and rapid offensive."*

As we've said, a majority of your time while campaigning will be focused on preparing to act; raising money, recruiting and training volunteers, gathering endorsements, etc. Then, in the final days of the election, when voters are paying the most attention, you should be able to rapidly mobilize your organization to outreach to as many of those unidentified persuadable voters as possible.

The graph of voter contacts over time is referred to as "the hockey stick" by some campaign staffers, because throughout the majority of the campaign your numbers will be quite low to the ground, and then will grow rapidly in the last few weeks and months, giving the resulting graph the appearance of a hockey stick. Once that rapid rise in activity comes, there will be no time to hire additional staff or recruit additional volunteers to fill in the gaps.

Thus, I cannot express enough the importance of taking into account the economy of action in your campaign plan. Take a look at each of the activities and habits you've outlined in your plan. Now ask yourself if there are any single activities that bring you closer to reaching two goals. Look for as many of these as you can and prioritize them.

In recent years, with the increase in mail-in ballots, we've been able to analyze much more accurately exactly when voters are making up their minds. Studies have shown that the most partisan voters turn in their ballots nearly as soon as they're able, while those in the persuadable universes are most likely to wait an additional few weeks. Thus, we can know fairly definitively that persuasion becomes more effective the closer it is to the election. But again, with the increase of early and absentee voting, the percentage of voters who've already returned their ballots will increase daily and your messaging will become less and less cost effective. If you wait too long, you'll mostly be talking to people who've already voted.

In voter contact, repetition is extremely valuable. A marketing principle developed in the 1930's with the advent of modern advertising describes this well. It's often called the "Rule of Seven" and says that someone needs to hear your message at least seven times before it sinks in. Think carefully about how you will reach each of your targets multiple times within such a short span of time. Without exponential growth, this will not be possible. So build your team strategically, employ their skills where they are most likely to be effective and conserve your strength for when it matters most.

Part 4: Adapting Your Strategy

" No plan survives first contact with the enemy."

Helmuth von Moltke the Elder

No matter how perfect your strategy, it will need to be adapted once your opponents make their moves. Despite what we would like, humans are not completely rational actors, and thus you will never be able to predict exactly what your opponents will do, or how voters will react to your message.

In Part 4: "Adapting your Strategy," we delve into the art of responsive campaign planning. Here we confront the reality that in the unpredictable world of politics, even the most meticulously crafted strategies need to be adaptable. While the idea of things not going to plan has an unhappy connotation, remember that change is not inherently negative.

This section of the book will equip you with the tools and mindset to turn obstacles into advantages. It emphasizes that a static strategy is a losing one. The ability to adapt is one of the keys to your campaign's success.

Chapter 6 discusses assessing your strengths and weaknesses, comparing them to your opponents', and how to use the understanding you have gained from this analysis to ensure your opponent only has access to attack the points where you are at your strongest. And conversely, how to attack only the areas where your opponent is at his weakest.

Chapters 7 and 8 delve into contingency planning and risk management—critical aspects of strategic planning that are often neglected. These chapters emphasize the importance of foreseeing potential challenges and preparing for them ahead of time. Here we will explore the value of being first in the field, ready for action, and the power of a proactive strategy to influence your opponent.

The second half of Chapter 10, which was moved here because it better aligns with the themes in this part of the book, delves into risk management, discussing the strategies and tools to anticipate, assess, and mitigate risks in a campaign context.

VI. Weak Points and Strong

He who can modify his tactics in relation to his opponent and thereby succeed in winning, may be called a heaven-born captain.

We've already discussed the importance of tailoring your strategy to the circumstances provided by the voters and the opponents you face. However, circumstances sometimes change, and on campaigns they can change fast.

Perhaps a new poll comes out showing the electorate has changed, or a scandal at the national level lights a fire under some of the voters in your district. This might increase or decrease the expected turnout on election day, or a hot-button issue might suddenly become a litmus test shifting a large section of the electorate towards one party or the other. Regardless of the specifics, if it changes the election day math, then your plans will need to change as well.

Whoever is first in the field and awaits the coming of the enemy, will be fresh for the fight; whoever is second in the field and has to hasten to battle will arrive exhausted. Therefore the clever combatant imposes his will on the enemy, but does not allow the enemy's will to be imposed on him.

The easiest way to profit is to be the first to seize the new opportunity. Being first in the field means being the first to release new policy positions or to engage with newly registered voters. If new polling shows that a previously one-sided district is suddenly in play, the first candidate to declare themselves is likely to be the one who will benefit. They will receive the support of party officials, donors, and volunteers just for showing up, while subsequent candidates will need to prove why they are more deserving.

By holding out advantages to him, he can cause the enemy to approach of his own accord; or, by inflicting damage, he can make it impossible for the enemy to draw near. If the enemy is taking his ease, he can harass him; if well supplied with food, he can starve him out; if quietly encamped, he can force him to move.

Remember the principle that all warfare is based on deception, and do your best to keep your opponents off-balance. They should be reacting to your moves instead of the other way around. In the previous chapter on Energy, Sun Tzu emphasizes the importance of applying pressure to keep opponents off balance. If something has changed, and your opponent has not yet been able to adapt, applying extra pressure will put them under additional strain, leaving them vulnerable and you strong.

For example, if an internal poll shows voters' opinions have changed, you may be able to bait your opponent into saying something stupid before they know their views are no longer popular. You might be able to use the new information you've learned to draw them into a debate on the position they haven't planned or prepared for.

> You can be sure of succeeding in your attacks if you only attack places which are undefended. You can ensure the safety of your defense if you only hold positions that cannot be attacked.

> Hence that general is skillful in attack whose opponent does not know what to defend; and he is skillful in defense whose opponent does not know what to attack.

Another strong example is targeting demographics or issues that your opponent has neglected. If you discover they have overlooked a specific community or haven't addressed a pressing issue, you can step in with a strong message and grab up the advantage for your campaign. This is all about finding and filling the voids left by the opposition.

Similarly beware your opponent catching you unaware in the same way. You must ensure that your information is up-to-date and your policy positions are well-researched. Give your opponent as few gaps to exploit as possible.

Remember that unpredictability is a strength. By championing an issue that has been flying under-the-radar or by forming an unexpected alliance, you can keep your opponent on the back foot, and renew or regain the first-mover advantage.

> By discovering the enemy's dispositions and remaining invisible ourselves, we can keep our forces concentrated, while the enemy's must be divided. Hence we shall be many to the enemy's few. And if we are able thus to attack an inferior force with a superior one, our opponents will be in dire straits.

This is the core principle of an adaptable strategy: Discover where your opponent is weak but you are strong and use this to your advantage.

But, to unleash the full power of an unexpected action, you must understand the psychology of your opponent—to know your enemy as you know yourself. Put yourself in their shoes and try to think like them. Dig into their past, uncover their vulnerabilities, research their weaknesses and strengths, and use this knowledge to predict their behavior as best you can. When a situation changes, and they become aware of it, how will they respond? By gaining an understanding of their psychology, you can exploit their biases and blind spots to capitalize on changes in circumstance they cannot even see.

This is much easier against incumbents and those who are well-known within the community as they are likely to be more predictable, and will have a position on record that they cannot escape.

Caitlin Huxley

The spot where we intend to fight must not be made known; for then the enemy will have to prepare against a possible attack at several different points; and his forces being thus distributed in many directions, the numbers we shall have to face at any given point will be proportionately few.

If you can hold the first mover advantage, if you truly understand your opponent, and if you're lucky enough to catch them in a trap, you may be able to split their attention enough to force them to fight on two fronts. This is the essence of Divide and Conquer, and is a tactic that has been used by leaders throughout history to gain the upper hand.

Again let's use the example of voter targeting; Obviously, you will win over voters from your party, and your opponent will win over theirs. But at the edges of the swing universe, there are Democrats and Republicans who cross party lines to vote for candidates they relate to more.

If a new issue changes the makeup of these universes, and you find a part of your base is suddenly in play for your opponent, would you change your plans? What if you discovered they are now targeting your supporters? If you decide to focus heavily on locking down this group, then imagine the shock and frustration you would feel if you discovered your opponent was not focusing on them, and the info you received was merely a feint, meant only to distract you. On the other hand, if you answered no, then consider the consequences of discovering that your opponent was making a real push for those swing voters, and you didn't defend against it.

This is how a change in circumstances can provide an

opportunity to divide a candidate's attention and leave them not knowing how to act. If you catch your opponent in a situation like described above, simply lit-dropping the neighborhood where those voters live once or twice can leave him thinking you're going to focus on it. If you are caught in the situation above, make very sure of the facts before you act.

> All men can see the tactics whereby I conquer, but what none can see is the strategy out of which victory is evolved. Do not repeat the tactics which have gained you one victory, but let your methods be regulated by the infinite circumstances.

In all cases, your campaigns must be prepared to innovate and experiment to stay ahead of the game. A great source of inspiration can be found after nearly every presidential election cycle, when the winning campaign staff puts out a book on how they changed the game, adapted their tactics, or came at their opponent from left field. Here you might find new ideas and tactics to employ in your race.

However, be careful. When you look at successful campaigns from the past, it's tempting to imitate the tactics you see. But the depth of research and planning that went into those decisions is often invisible. You almost certainly face a different situation than they did; You face a different opponent if nothing else. Every campaign is unique and requires its own tailored plan.

Sun Tzu's wisdom is particularly relevant here: "All men can see the tactics whereby I conquer, but what none can see is the strategy out of which victory is evolved." Don't simply repeat what has worked in the past, but let your methods be shaped by the ever-changing circumstances.

Above all, remember that the path to victory isn't a straight path. It's a dynamic journey, and you must constantly be adapting to change and using the new circumstances to outmaneuver your opponent. This is your key to a winning campaign, so stay flexible and focused.

VII.
Maneuvering

In war, the general receives his commands from the sovereign. Having collected an army, he must blend and harmonize the different elements thereof.

After that, comes tactical maneuvering, than which there is nothing more difficult. The difficulty of tactical maneuvering consists in turning the devious into the direct, and misfortune into gain.

In any campaign, the candidate sets the agenda. It then falls to the campaign manager to ensure that all facets of the operation (communications, fundraising, volunteer outreach, etc.) are in sync and working smoothly together. This alignment forms the solid foundation upon which the campaign can gather momentum.

However, it's important to recognize that this foundation, while necessary, is not enough. Winning goes beyond merely assembling a competent team and adhering rigidly to a strategy. The unpredictable nature of elections demands that you be prepared to adapt and a willingness to take unanticipated paths. Maneuvering can entail re-deploying resources to exploit an unexpected opportunity.

While the previous chapter explored responding to the truly unexpected, this chapter underscores the importance of contingency plans. This is where you imagine the possible scenarios of what things might go wrong, and how you will capitalize on them if and when they do; turning misfortune into gain, as Sun Tzu says.

Consider the possibilities of a major donor or endorsement pulling out, an important staffer quitting, or new polling showing a major shift in voter sentiment. None of these issues are foreseeable, but all are quite common. After you've outlined your responses for these frequent scenarios, keep going; continue to brainstorm less-obvious challenges and how you'd handle them. Do this exercise with your campaign manager and close advisors to benefit from the greatest depth of experience.

Maneuvering with an army is advantageous; with an undisciplined multitude, most dangerous.

If you set a fully equipped army in march to snatch an advantage, the chances are that you will be too late. On the other hand, to detach a flying column for the purpose involves the sacrifice of its baggage and stores.

Just be careful. Adaptability means being prepared to make smart, calculated shifts that align with your overall goals—not abandoning your core strategy. Success here is about being nimble without losing sight of your goals. In fact, there is often not enough time to completely reevaluate your plan, re-train or re-deploy your team, or make all of the other preparations you would like.

Consider the hypothetical of new polling showing that an unexpected shift in voter sentiment has suddenly made a neighborhood competitive. If you take the time to build up your presence in this area the way you have for the other targeted areas of your district, you're likely to miss your window of opportunity. On the other hand, if you respond rapidly with just whatever resources you have at hand, that would mean separating your team from their support system, and stretching them too thin to ensure you can seize the advantage.

The key here lies in preparation and planning; having extra room in your plans for these scenarios, in the same way as you would add a miscellaneous buffer to your budget. This is how you can position yourself to be able to adapt swiftly without sacrificing the support system your team needs.

> We cannot enter into alliances until we are acquainted with the designs of our neighbors. We are not fit to lead an army on the march unless we are familiar with the face of the country—its mountains and forests, its pitfalls and precipices. We shall be unable to turn natural advantage to account unless we make use of local guides.

When planning, don't overlook the reactions of your allies and volunteers. Understand their goals to anticipate their responses to each of the different scenarios you've outlined. You can mostly control your staff, but outside influences can shift the dynamics. Think about how unexpected changes like a surprise endorsement or the loss of a key supporter might ripple through your campaign. Remember, you're not the great unifier; your allies have their own networks and conflicts, and your plans should take this into account.

In the suburbs of Chicago, my candidate once faced a situation where there were two feuding local Republican groups, each with loyal donors and volunteers to help the candidates they endorsed. The issue was that neither would work with candidates endorsed by the other. We had to court one of them, but how to choose? Unfortunately, in situations like this, there is often no right answer, and attempting to please both sides would have risked alienating everyone. So we did the only thing we could do. We laid out two sets of plans, each one aligning with a different group, considering their impact on our campaign. We listed the pros and cons, and thought how they would react in each of our contingency scenarios, then selected the one that best served our needs.

Recognize that you can't know every group and individual in your district or how each will respond. Instead engage local guides who have a closer understanding of the dynamics to help you refine your plans, providing insights that enable you to anticipate reactions and tailor your approach to fit the community.

This research and planning should provide you with a detailed map of your district and the voters within it. When a new situation arises, you'll be able to consider how it will affect each voter segment, and maneuver your campaign into the best position to seize the advantage.

In war, practice dissimulation, and you will succeed. Whether to concentrate or to divide your troops, must be decided by circumstances.

Once your plans are set, and contingencies are decided, the difficulty comes in ensuring that the right people on your team get their orders without the plans becoming public knowledge. In the example above with the two rival Republican groups, it was vital that our list of pros and cons remaine private. In your campaign, you will need to be careful to keep your strategy secure.

When you plunder a countryside, let the spoil be divided amongst your men; when you capture new territory, cut it up into allotments for the benefit of the soldiery.

Remember that victory won't come merely from meticulous planning; it also comes from the hard work, dedication, and sometimes exceptional achievements of your staff and volunteers. Consider now how you will reward them when something exceptional happens. What if they reach, or even exceed, their goals? What if someone in the field brings in a new large donor or secures the support of a group you didn't even know about?

This may go against your initial instinct, as campaigns are often run on a tight budget and managing resources can be difficult, but it is important to invest in your staff and have plans in place to keep them motivated and engaged, or your campaign will pay for it in other ways. Plan for these successes, and do your best to encourage excellence and celebrate accomplishment.

The Book of Army Management says: On the field of battle, the spoken word does not carry far enough: hence the institution of gongs and drums. Nor can ordinary objects be seen clearly enough: hence the institution of banners and flags.

On your campaign, the flow of communication must be two-way, allowing for real-time feedback. Team members on the ground are your eyes and ears, offering immediate insight into voter responses, gauging shifts in the political climate, and identifying emerging opportunities or threats. In this way, you can adapt your plans as the situation on the ground changes, allowing for real-time response to unexpected events.

With modern technologies, changes in strategy can be rolled out nearly instantly, but these updates must be presented in an easily digestible format. A new tactic is only effective if everyone on your team understands it and can apply it correctly. The tool is only as effective as your ability to utilize it.

While the insights from your field team are invaluable for gauging voter sentiment, do not get so immersed in the details that you lose sight of your strategy. "Listening to the grass-roots" should inform your tactics, but not dictate them. Be careful not to become a campaign that's always reacting, chasing the latest trend or concern, and in doing so, missing the forest for the trees. Maintain your strategic focus, balancing immediate feedback with long-term goals and overarching campaign objectives. This will help ensure that your contingency plans allow for adaptability without causing you to lose sight of the broader vision.

Now a soldier's spirit is keenest in the morning; by noonday it has begun to flag; and in the evening, his mind is bent only on returning to camp. A clever general, therefore, avoids an army when its spirit is keen, but attacks it when it is sluggish and inclined to return.

When I am in the field, I know my volunteers will become sluggish as the day goes on, and so when it's time to break for lunch, I change tack. If we go back to the office, we can eat and talk while we stuff envelopes. If we go to a bar or restaurant, then we do it as a group, so the volunteers can build camaraderie and a sense of community. Having this plan in your back pocket is much more effective than trying to adapt in the middle of the day when your volunteers start peeling off.

Likewise, at the beginning of a campaign, volunteers and staff are filled with enthusiasm and energy. They are eager to get out there and make a difference. But as the days turn into weeks and months, that excitement tends to wear down. Burnout is a campaign killer, and I have seen firsthand how it can drain the life out of a campaign, sapping the energy of volunteers and donors alike. If you see it beginning to set in on your team, you'll need to react quickly—which means you need to have a change of course prepared ahead of time.

Contingency planning involves not only big strategic decisions but also daily adaptability, recognizing the natural ebb and flow of energy within your team. To harness this, you should plan engaging activities for when energy is highest, but pivot when needed for simpler tasks, allowing for rest and recuperation.

To be near the goal while the enemy is still far from it, to wait at ease while the enemy is toiling and struggling, to be well-fed while the enemy is famished—this is the art of husbanding one's strength.

To refrain from intercepting an enemy whose banners are in perfect order—this is the art of studying circumstances.

Mastering the art of timing is not just about playing to your team's strengths but also about exploiting your opponent's vulnerabilities. That is why it is so important not to focus only on your plans, your allies, and your team, but also the circumstances faced by your opponents. If you understand their weaknesses and vulnerabilities, you can create plans that will exploit these gaps if they should start to widen. Similarly, by making your plans adaptable, you can pivot away from an engagement if for example, you realize your opponent is not as weak as you thought and you no longer think you will fare so well. In this way, your campaign can conserve its resources and avoid overexertion, while the opposition is left struggling to keep up.

A great example of this can be seen in the 2008 Democratic Primary between not-yet-president Obama, and Hillary Clinton, who was widely seen as the frontrunner. The Obama campaign focused their efforts on the caucus states, where grassroots engagement would make the greatest impact. Obama's targeted approach caught his competitor off-guard, and secured him the nomination. By the time the Clinton campaign realized what was going on, it was too late.

It is a military axiom not to advance uphill against the enemy, nor to oppose him when he comes downhill.

- *Do not pursue an enemy who simulates flight;*
- *Do not attack soldiers whose temper is keen.*
- *Do not swallow bait offered by the enemy.*
- *Do not interfere with an army that is returning home.*
- *When you surround an army, leave an outlet free. Do not press a desperate foe too hard.*

Such is the art of warfare.

Do not make the mistake of underestimating your opponent. Just as an army should not advance uphill against an enemy, your campaign should not take on an opponent who is currently in a position of strength. Earlier I said that you should not stray too far from your original strategy. But, if circumstances have changed drastically, and your opponent is now favored to win, your strategy will need to change as well. This helps you pivot effectively when your opponent gains an upper hand.

Every campaign strategy carries its own risks, and your contingency plans are your safety nets against polling downturns, scandal, or sudden funding losses. Keep pushing for additional advantages, but avoid the temptation to overextend, which could leave you vulnerable to financial strain or running out of time to accomplish everything. And if you find your opponent in a weak moment, such as a gaffe or a controversy, it's sometimes wiser to step back and allow public opinion to take its course. Give them enough rope and they may hang themself.

In one of my early campaigns, we recognized the growing influence of social media and decided to innovate. We assembled a "Social Media Strike Team," composed of supporters who were active on social media and had significant followings of their own. We could create a social media post and by sharing it with our strike team, the message could rapidly spread through their own networks, amplifying our reach. This strategy paid off when our opponent launched a surprise negative ad campaign against us, and we were able to release our message quickly, and shape the narrative.

This is the kind of flexible contingency that can be woven into your campaign strategy from the beginning. It's not just about having a plan, but having a plan to adapt. Remember that a good campaign knows what to do when things go right; but a great campaign knows what to do when things go wrong.

VIII. Variation in Tactics

To take a long and circuitous route, after enticing the enemy out of the way, and though starting after him, to contrive to reach the goal before him, shows knowledge of the artifice of deviation.

Each strategy or tactic you employ will carry with it some risks. It might fail to have the desired impact or may backfire completely. Effective risk management lies in recognizing the variables, making informed decisions, and taking steps to mitigate the risks; knowing when to advance beyond the mark or when to pull back, when to stick to the plan, and when to deviate. Many of these are decisions that must be made in the heat of the moment, but without proper training and experience, it will be difficult to know which is which. Still, there are some best practices to follow.

To win, campaigns must be agile and responsive, always ready to respond to new developments and shifting public opinion.

In war, the general receives his commands from the sovereign, collects his army and concentrates his forces.

When in difficult country, do not encamp. In country where high roads intersect, join with your allies. Do not linger in isolated positions. In hemmed-in situations, you must resort to stratagem. In desperate position, you must fight.

There are roads which must not be followed, armies which must be not attacked, towns which must be besieged, positions which must not be contested, commands of the sovereign which must not be obeyed.

Sun Tzu's warning reminds us of the importance of making your decisions based on careful analysis and evaluation. In the complex political landscape, you must be extra mindful of the potential consequences of your actions. When you are in a tough spot, maybe under attack or being criticized on an issue, do not just linger. Instead, seek out opportunities for alliances and partnerships. By joining hands with like-minded allies, you can gain a strategic advantage and increase your chances of success.

If it feels like the walls are crumbling down around you, and your plans are falling apart, you'll need to get creative. If you have no other options, and your allies have abandoned you, you have no choice but to fight.

Candidates, cover your ears, because this message is just for campaign managers and staffers: You may be given orders by your candidate that are not in line with what is best for the campaign. In these situations, use your expertise and knowledge of the district to advise your candidate on the right course of action and above all, avoid implementing strategies that you are are likely to fail. Candidates, like sovereigns of old, do not always know what is best, and only a fool would try to follow a strategy he knows will fail.

If you are a candidate, then for your part, you should not give orders out of hubris or malice and should listen when your staff and advisors caution you about a decision.

> *Reduce the hostile chiefs by inflicting damage on them; and make trouble for them, and keep them constantly engaged; hold out specious allurements, and make them rush to any given point.*

Keep your opponents and their supporters busy as best you can. It's not enough just to provide your opponents with distractions. To succeed, you must also find ways to disrupt the networks and relationships that support them. This means targeting their donors, volunteers, and coalition leaders, and finding ways to undermine their ability to mobilize and organize. Create divisions within their ranks by playing on the different motivations and interests of the groups within their coalition. This way, you can sow seeds of discord and prevent them from coming together in a united front.

Additionally, Many states require candidates to gather petition signatures to secure a spot on the ballot. These signatures are open to legal challenges, offering a strategic avenue to keep your opponent preoccupied and financially strained. Clever strategists know how to use all the tools at their disposal.

The art of war teaches us to rely not on the likelihood of the enemy's not coming, but on our own readiness to receive him; not on the chance of his not attacking, but rather on the fact that we have made our position unassailable.

You should be proactive and prepared for any potential opponent. Rather than hoping someone won't run or for a weak and ineffective opponent, you must take active steps to put yourself in a position of strength and to distract and disrupt any would-be opponents' efforts at getting established. You must focus on your strengths, and work to shore up any weaknesses in your campaign. This means having a strong message and platform, a well-organized team, and a clear strategy for victory. Remember that *"The good fighters of old first put themselves beyond the possibility of defeat..."* By doing this, you can be confident and ready to face any opponent, no matter how formidable they may seem, and by making your position unassailable, you can ensure that you are in the best possible position to win.

One of the easiest ways to make your position unassailable as an incumbent, is by constantly focusing on your campaign's potential energy, as outlined in Chapter 4. Grow your coalitions, train your volunteers, and ensure your organization is constantly in a state of readiness. As the Roman military writer Vegetius put it, *'Si vis pacem, para bellum'*—*'If you want peace, prepare for war.'* By keeping your team trained and strong, you're not just building a team that's ready for the campaign ahead, not defending your position and making a campaign less likely—only a fool would challenge an incumbent as strong as you.

There are five dangerous faults which may affect a general:
(1) Recklessness, which leads to destruction.
(2) Cowardice, which leads to capture.
(3) A hasty temper, which can be provoked by insults.
(4) A delicacy of honor which is sensitive to shame.
(5) Over-solicitude for his men, which exposes him to worry and trouble.

Recklessness occurs when the heat of the campaign is at its highest, and candidates are pressured to make decisions they don't feel ready for. This often occurs in smaller races, where the candidate was not recruited until later in the campaign cycle and sees their opponents already out knocking on doors, their yard signs popping up all around the district. In response, these candidates may put the cart before the horse, knocking on doors without adequate literature or without first recruiting volunteers to help them grow their campaign. To combat this, you should learn the concept of "Sharpening the Saw" from Stephen Covey's *The Seven Habits of Highly Effective People*. Just as a tree cannot be cut down with a dull saw, voters cannot be effectively persuaded without a well-organized campaign in place to contact them.

Cowardice is much easier to see. Many canvassers struggle with the fear of rejection or the discomfort of asking sensitive questions. It may be awkward and uncomfortable to ask for someone's email or press them into telling you who they will vote for in an upcoming election. But it is going to be much more uncomfortable when you can't ask your supporters to volunteer because you don't know who they are. A fix for this issue is to "begin with the end in mind."

Before even setting out to knock on doors, set a clear goal in your mind and calculate how each action you take will contribute to achieving that goal. By focusing on the end goal, training staff to deliver your message with confidence, and providing ongoing support and guidance, campaign managers can help their teams to overcome their fears and succeed in their outreach efforts.

A hasty temper has already been touched upon in Chapter 1, *"If your opponent is of choleric temper, seek to irritate him"*. This applies just the same when you're on the defensive, as your reputation is likely to be attacked by your opponents. They may resort to stretching the truth, or even outright lies, to damage your standing in the eyes of the public. In such situations, do your best to remain calm and composed. Before you respond or take any action, consider how your opponent would like you to react, and what the most advantageous course of action would be from an objective perspective. By thinking like an outside observer, you'll be better equipped to come up with a plan of action that will minimize the damage to your reputation and maximize your chances of success.

Similarly, shame is a feeling we all feel from time to time, and when candidates face setbacks or attacks in the public sphere, they may feel ashamed. If this happens to you, you might feel tempted to do what you perceive as "the right thing" rather than what is advantageous for your campaign. But, this often stems from a fear of backlash from voters or a reluctance to take risks.

The last fault, over-solicitude for my men, is the one I find the hardest to avoid. I try to foster a sense of community and belonging among my staff, interns, and volunteers because when they feel like they are part of a family, they will be more motivated to work hard and fight for the success of the campaign. However, be cautious not to let this sense of familiarity and comfort go too far. If your staff becomes too comfortable and relaxed in the campaign headquarters, they may lose their professional edge and work ethic. It is crucial to strike a balance between friendliness and professionalism.

> *When an army is overthrown and its leader slain, the cause will surely be found among these five dangerous faults. Let them be a subject of meditation.*

Winning an election is no easy feat. It takes hard work, dedication, and strategic planning to emerge victorious. But even the best-laid plans can be foiled by the presence of certain dangerous personality traits in a candidate or campaign manager.

Recklessness, or a lack of caution and foresight, can lead to rash decisions that ultimately result in destruction. Cowardice can be just as detrimental, and a candidate or campaign manager who lacks the courage to stand up for their beliefs or to do what they need to do will watch all their advantages be captured by their opponents easily. Insults and criticism are par for the course in politics, and a candidate or campaign manager with a hasty temper, who cannot handle these without losing their cool will struggle to maintain their composure under the pressure of the campaign. A sense of honor that is easily offended by perceived shame can be a liability in politics. Being overly sensitive to criticism will leave you struggling to maintain composure, and has caused more than one candidate to quit mid-cycle. While it is important to care about and support one's team, over-solicitude for them rapidly destroys the professionalism needed to keep the team working hard until election day.

These are by far, the most dangerous personality traits in a candidate or campaign manager, which can negatively and severely affect your chances of winning your election. In every campaign that by all accounts should have won but somehow didn't manage to get there, there was undoubtedly some combination of these traits present in the leadership.

When I first began consulting, I was involved in a primary where two of our primary opponents got so fixated on a personal feud that between the two of them, that they lost sight of the real frontrunner, my candidate. Guided by recklessness, a hasty temper, and a sensitive sense of honor, they squandered time and resources on this grudge. The result? Both feuding candidates lost, and my candidate slipped by unnoticed.

Stories like this are commonplace, and most campaign staffers can recount multiple instances of promising campaigns they've witnessed derailed by a faulty general.

X. Terrain: Section 2

Now an army is exposed to six several calamities, not arising from natural causes, but from faults for which the general is responsible. These are:

(1) Flight, (2) insubordination, (3) collapse, (4) ruin, (5) disorganization, and (6) rout.

These are six ways of courting defeat, which must be carefully noted by the general who has attained a responsible post.

In response to conflicts you encounter, think of these six types of misfortunes you might inadvertently invite into your camp, and ask yourself if you are in danger from any of them.

Other conditions being equal, if one force is hurled against another ten times its size, the result will be the flight of the former.

Flight: If the voters within the district are split 90 to 10 on an issue, and you align yourself with the weaker position, even your party and volunteers will be unable to support you. You will find yourself completely alone. This sounds incredibly obvious but it's also incredibly common. If you feel strongly about something, and the incumbent in your home district stands in direct opposition, you might think that you need to run against him. Without this insight, you might feel pressed to make a stand. If you do, you will lose.

When I was executive director for the Chicago Republican Party, we had candidates popping up all across the city wanting to run against their Democratic state rep or alderman (that's like a city-council member, for you non-Chicagoans). They didn't think about the makeup of the district, or their chances for success before declaring their candidacy. They just saw an elected official who didn't represent their views, who had been without a serious challenger in years. Aiming to change that, they put together what resources they could, and began building their campaigns. Because the local party stopped and did the math, these candidates received little to no support, and after their inevitable loss, were never heard from in politics again. It is my opinion that a dose of the truth would save many capable and willing candidates from burning out and fading away in this way.

When the common soldiers are too strong and their officers too weak, the result is insubordination.

Insubordination: If your staff are too inexperienced to keep volunteers on the script, or if your volunteers have strong opinions on irrelevant issues, there will be insubordination. A strong training program for both staff and volunteers is incredibly important for this reason. Keep a good line of communication about the purpose of each activity, and make sure everyone has a clear understanding of how it contributes to the overall victory of your campaign. Every campaign has this problem to some degree.

The voter identification phase makes up one of the largest portions of the campaign cycle. It's purpose is to gather accurate data, hone your target list, and identify supporters who might donate or become volunteers themselves. Inevitably a volunteer will misunderstand the point of this activity, and spend time arguing about issues with people who might be potential supporters. If they try to sway people instead of just identifying them, they could inflate your supporter list with voters who shouldn't be there. This leads to misdirected get-out-the-vote (GOTV) efforts, as you spin your wheels targeting the wrong people. The result is wasted time and resources on non-supporters.

When the officers are too strong and the common soldiers too weak, the result is collapse.

Collapse: If your paid staff are too forceful, and your volunteers have not been trained, they will be embarrassed and frustrated, and your organization will break down for lack of return volunteers.

New volunteers are often intimidated by the prospect of talking to voters, and if pressed too quickly outside their comfort zone, they may opt to simply go home instead, and that is probably the best outcome. If they are forced to do a job for which they are unprepared, they may do even more harm to your campaign.

This is why good campaign staffers build complex volunteer funnels; to move people from one step to the next at a pace they are comfortable with, all the while increasing their workload or responsibilities. Slow and steady wins the race, but sometimes campaigns are short on time. If you must work quickly to build your organization, and if positive experiences aren't built on top of other good experiences, volunteers will tire out before election day. When it's time for GOTV, they will be exhausted and their loyalty waning, they will desert you.

When the higher officers are angry and insubordinate, and on meeting the enemy give battle on their own account from a feeling of resentment, before the commander-in-chief can tell whether or not he is in a position to fight, the result is ruin.

Ruin: When your team, whether staff or volunteers, is not adequately disciplined, they may act in ways that can potentially harm your campaign. If you cannot ensure that your team is working in sync with the central campaign strategy, you will find yourself in a chaotic situation akin to herding cats—futile and inefficient.

This is a common issue, particularly in campaigns that rely heavily on volunteers. Without the incentives of previous experience or a paycheck, maintaining control and guiding their actions become daunting tasks. Thus, it is essential to identify alternate sources of motivation and ensure that your staff, especially, feels distinguished from your volunteer force. If they are not on board with the campaign strategy or they disagree with your plan, you may encounter significant challenges.

When the general is weak and without authority; when his orders are not clear and distinct; when there are no fixed duties assigned to officers and men, and the ranks are formed in a slovenly haphazard manner, the result is utter disorganization.

Disorganization: Your paid staff must be capable enough to manage your volunteers effectively without your direct involvement, and the orders coming from the top must be firm and forceful enough to drive your staff. If a campaign manager is inexperienced, gives unclear or confusing orders, or doesn't manage the paid staffers well, you will have an "organization" in name only. If your paid staff believe in the campaign, in the plan, and the campaign manager, they will have no issue following orders

When I have an office manager or a volunteer coordinator who is not performing as well as I would like, this is how I choose to fix the problem. Sitting down with them, I make sure they understand how the work they are doing contributes to the overall campaign plan. They should be able to see a clear link between their performance at the role, your success, and their likelihood to receive a win bonus. I also try to remember that my role as a manager is to help my staff be successful. So, if they can be retrained, or if they need an assistant, or need me to spend a few days with them in the field, then I try to be attentive to that.

When a general, unable to estimate the enemy's strength, allows an inferior force to engage a larger one, or hurls a weak detachment against a powerful one, and neglects to place picked soldiers in the front rank, the result must be rout.

Rout: If you have picked your race poorly, your opponent is well-staffed and funded, or you are unprepared to take them on, you will have no chance of winning. You must accept this.

When I teach new and inexperienced candidates, staffers, or elected officials, and I get to the part about estimating your chances, I am invariably asked what they should do if their inherent vote deficit is greater than the number of swing voters in their district. This is a tough question; not because it's hard to answer, but because the answer is likely to upset them. If this is you, don't run for that race, unless a tactical loss is part of your plan.

When we get to calculating the number of volunteer hours needed to contact all the target voters in a chosen universe, I'm often asked "What if there are fewer days until election day than it says I will need?". If this is you, look to the election after next, and run your calculations for the longer time frame.

Future staffers often have an easier time understanding these sorts of rules than candidates and elected officials, because they are likely to be a step or two further from the consequences, and thus be more strategy focused. When you are the candidate, you've got a lot riding on this particular election and it's going to be hard to accept that you can't win, or that the hated incumbent will remain in office for another 2–4 years.

Part 5: Navigating the Political Landscape

"The only real voyage of discovery consists not in seeking new landscapes but in having new eyes"

Marcel Proust

We often interpret the world through the lens of our assumptions. This can interfere with our ability to understand the diverse perspectives held by voters, opponents, and allies. In Part 5, we will challenge these assumptions and discuss how to navigate the political landscape by "having new eyes," as Proust calls it. This means stepping outside our comfort zones and seeking to understand the terrain from multiple angles.

In Chapter 9: "The Army on the March," we will explore strategies to navigate the political terrain effectively, in terms of your physical position within the district and your positions on issues. In the first half of Chapter 10: "Terrain," we delve into the possible characteristics of your district and how understanding its disposition can shape your campaign strategy. Chapter 11: "The Nine Situations," examines a spectrum of circumstances you might find yourself in, and provides best practices for responding to each.

As you read through this section, keep in mind the value of openness and continuous learning to best respond to the dynamic and often unpredictable nature of campaigns. I hope that, as you venture into this part of your journey you will find new ways of seeing that enhance your strategic prowess.

IX. The Army on the March

We come now to the question of encamping the army, and observing signs of the enemy.

Up to now, we've discussed only briefly the importance of strategic positioning. Research on your district is important to know where to target your efforts, but many new candidates and campaign staffers think only in terms of how they will maneuver while actively campaigning, and overlook how their initial positioning can influence voter perception, resource allocation, and overall campaign momentum. To avoid this, you should ask yourself: Which position can you put yourself in so that volunteers approach you, so that voters decide to support you without needing to be contacted, etc? Additionally, in what ways will your opponent be doing the same, and how can you strategically position yourself to minimize his gains and maximize your own?

Pass quickly over mountains, and keep in the neighborhood of valleys. After crossing a river, you should get far away from it. In crossing salt-marshes, your sole concern should be to get over them quickly, without any delay. In dry, level country, take up an easily accessible position with rising ground to your right and on your rear, so that the danger may be in front, and safety lie behind.

These are the four useful branches of military knowledge which enabled the Yellow Emperor to vanquish four sovereigns.

At this stage, you should consider the entire course of your campaign. Imagine the types of positions you and your team may find yourselves in, and how you might find an advantage in each.

Consider the placement of your campaign offices & the events you hold, as well as your stance on the issues that are popular in your district. While these may seem like separate facets of a campaign, they are intrinsically linked by a set of guiding principles. The location you choose for your offices and events is a physical demonstration of your commitment to the communities and issues those areas of your district represent. In the same way, the policy positions you take now serve as a signal of your campaign priorities and a promise to voters who share your focus.

Think of what the mountains, rivers, marshes, and flat country represent in your district and in your election. In the same way that military leaders use geography to inform strategy, your campaign should use the unique features and concerns of your district as a guide.

You have to decide where to place your campaign offices; how to balance having the best access to the train and bus, located close to a freeway offramp, or otherwise being easily accessible to your would-be volunteers. Think about how they will get to your office from their homes and to the neighborhoods you'll be walking. Putting an office in your primary target area can be an easy way to kick off your walk days, and is much better than having to meet at a coffee shop. If your district is difficult to walk because it is very rural or very urban, this is less of a consideration.

Generally, it is best to open your offices, plan your events, and focus your door-knocking in your target areas. If your opponent has offices near an area of voters that both campaigns are targeting, you have a few options. Placing your office near theirs invites conflict. Placing your office on the opposing side avoids conflict. Opening offices in the middle allows you to increase the efficiency of reaching your targets but limits your access to other areas. If you have multiple target neighborhoods nearby, all full of high-priority voters, consider whether you should have one headquarters in between them all or multiple offices throughout.

When it comes to actually walking door-to-door, the terrain affects each district differently. In the field, you will have to deal with apartment buildings, farmland, suburban housing developments, and the like. When doing voter outreach to an area, and you come across houses that are otherwise inaccessible; because they live in gated communities, in apartment complexes, on farms, or in other extremely rural housing, you will need to find another way in. Think about the voter makeup of the neighborhoods you're targeting, and whether it might make sense to have an office nearby.

In my campaigns, I always recommend a data-centric approach to field office placement. Prioritize swing areas of the district where your policy stances resonate most strongly with the local electorate. This ensures efficient resource use, targeted outreach, and strengthens community ties. The strategic location of your offices amplifies your campaign's commitment to the community, fortifying relationships with volunteers and target voters.

Similarly, in the realm of issue positions, you must make sure that your campaign is within range of the majority of voters. When planning your position on policy issues, the best advice comes from game theory and is known as the median voter theorem (aka Hotelling's game). The basic explanation is that voters will vote for the candidate who is closest to them on the political spectrum and that candidates will, over time, drift toward the center. Take advantage of this law of human nature and put yourself between your opponent and the bulk of the voters; just don't get so close that voters can't tell you apart. If your opponent has an unpopular stance on an issue, position yourself on the more favorable side of the issue, then highlight it to voters! If they have a popular opinion, agree with them, and quibble only on the details. This goes doubly for cross-party issues; If you are a conservative, running in a district with a heavy liberal lean on one issue, consider positioning yourself as a centrist on that issue.

These are the three major positional issues you will need to take into consideration. Make the right choices in each, and you will find yourself with a natural advantage in all the major parts of your election.

If you are careful of your men, and camp on hard ground, the army will be free from disease of every kind, and this will spell victory.

Consider the position that is best for your team. Failing to do so can result in a major disadvantage for your campaign.

An office that is located too far from your volunteers, or from the areas you intend to walk will result in it being used as a storage unit, and nothing more. If staff are required to go there regularly, the added travel time will contribute to fatigue. If volunteer events are held there, attendance will suffer, and as a result, your outreach counts will decrease, adding extra pressure on your staff to overcome the difference, as mentioned before.

If you plan walks in areas of the district where the voters' actual doors are difficult to reach, like in a very rural or urban area, your volunteers will not feel effective or like they are making a difference. Likewise, your staff's outreach numbers will suffer and they will feel the added pressure. If you do not equip them with the tools they will need to efficiently reach the voters in your district, they will be unable to spread your message, identify supporters, or persuade undecideds.

Likewise, if your stance on an issue puts you further from the bulk of your likely voters than your opponent, you will naturally lose a large portion of them. Remember that your volunteers and staffers are voters as well, and most will only work for candidates with whom they align. Once again, when your recruitment efforts suffer, your outreach capabilities will likewise suffer, and your campaign will be less likely to win. By ensuring your team has a solid foundation of respect and a shared purpose, you can safeguard your campaign against the epidemic of conflict and discontentment.

Country in which there are precipitous cliffs, tangled thickets, quagmires and crevasses, should be left with all possible speed and not approached. While we keep away from such places, we should get the enemy to approach them.

Just as you should work to position yourself to take as many advantages for yourself as you can, so too should you work to force your opponent into disadvantageous positions. While you cannot force him to sign a lease in the wrong part of the district or to walk in areas that are not easy to walk, you can use the research you have done to force him to highlight his positions on issues you know will be unpopular with his supporters and volunteers.

If in the neighborhood of your camp there should be any hilly country, or woods with thick undergrowth, they must be carefully routed out and searched; for these are places where men in ambush are likely to be lurking.

Remember again, that your opponent will have access to much of the same information you will. Everything you consider in your planning and positioning, he will be considering as well. If there are situations where you would be in a disadvantageous spot, you must make sure your opponent is not in a position to make good use of that fact. If you are considering placing your office in a remote corner of your district, first ensure that your opponents are not placing theirs in your target areas. Before making any issue a major part of your campaign, you must be sure your opponent's position will not be more closely aligned with the district than yours.

> *When the enemy is close at hand, he is relying on the strength of his position. When he tries to provoke a battle, he is anxious for the other side to advance. If his place of encampment is easy of access, he is tendering a bait. Movement amongst the trees shows that the enemy is advancing.*

> *When envoys are sent with compliments in their mouths, it is a sign that the enemy wishes for a truce. What we can do is simply to concentrate all our available strength, keep a close watch on the enemy, and obtain reinforcements.*

Here Sun Tzu talks about looking for signs of your opponents positions. Pay close attention to everything your opponent says and does, especially when comparing/contrasting himself with you. If your opponent is calling for a debate, he believes he has outmaneuvered you, and perhaps thinks that he can paint you in a bad light over some issue. Look at what he has been saying lately, and make very sure that you are stronger on this issue with the bulk of likely voters in your district. Watch out for baits your opponent puts out for you. If he asks you a softball question, it's likely to be a trap.

This positional game is only possible if you have done your research, and you know your opponents' positions intimately.

Subscribe to their newsletters, follow them on social media, and press your staff to do the same. If they call for volunteers, consider sending an intern or staffer to collect materials and come back. Take steps to keep your opponents from being able to do the same. Again, more on this is written in Chapter 13: "The Use of Spies."

As I've said in a previous chapter, your name will most likely be dragged through the mud, probably with something you consider to be an obvious lie. You may be driven to respond, but first think to yourself that this may be bait, and consider your options. Beware of the Streisand effect, where trying to make an article or a criticism disappear ironically gives it more attention, and do not fan the flames. If you are less popular on an issue, in a neighborhood, or with a specific demographic, do not be the one to engage there.

You can also use information gathering to backward engineer his plan. Do what you can to observe his team in the field. If they drop literature in an area but do not knock on the doors to talk to the voters, this area may only be a secondary target. Think about the demographics of voters that live here, and what they might have in common. If he hosts large, well-funded, and well-coordinated walk days, then he has enough staff to manage them. But if he has volunteers scattered throughout the district, or is more often seen waving yard signs than knocking on doors, then he lacks a strong management structure. This information might come from his email blasts, from your spies showing up to his volunteer day, or from reviewing his campaign expenditures at the end of each filing period.

He who exercises no forethought but makes light of his opponents is sure to be captured by them.

In every situation think about the circumstances, what you consider to be your best options, and what your opponent would like you to do. If you do not make your plans with all these factors in mind, you risk missing something that could end up handing the advantage to your opponent. Similarly, underestimating your opponents is one of the deadliest and most common mistakes I see made by novice politicos. Be warned that it is one thing to disagree with your opponent's policy positions, and another to dismiss their strategic capabilities.

Train your staff to think in this way as well. A team that's in sync on this will be far more effective, and resilient against challenges.

* * *

Gaining every possible advantage may seem important, but not all opportunities are made equal, and you will often have to make difficult decisions. Do you put your office close to your target walk areas or use the party HQ that is offered for free? What do you do about an issue that is a deal-breaker to one of your allies, but does not align with voters in your district? If there is no true win-win scenario available, then go instead for the lesser of two evils.

X. Terrain: Section 1

We may distinguish six kinds of terrain, to wit:

(1) Accessible ground; (2) entangling ground; (3) temporizing ground; (4) narrow passes; (5) precipitous heights; (6) positions at a great distance from the enemy.

These six are the principles connected with Earth. The general who has attained a responsible post must be careful to study them.

Sun Tzu uses the idea of Ground to explain the pre-existing conditions in the district. For our purposes that is the most applicable to the positions on the issues held by voters in your district. The voters of the district will have opinions. Some stances you might take will be popular, others less so.

Use this list as a thought exercise. Think about the positions you might take on issues that matter to your district. Think about the six kinds of terrain, and which each might metaphorically represent. Consider the great generals' advice on the subject and whether it might be relevant to you.

Ground which can be freely traversed by both sides is called accessible. With regard to ground of this nature, be before the enemy in occupying the raised and sunny spots, and carefully guard your line of supplies. Then you will be able to fight with advantage.

Accessible Ground: An issue that can easily be held by both sides, such as a bi-partisan issue that everyone in the district supports. If you and your opponent both hold the popular opinion on an issue, then be the first one to declare yourself, so you can take the first-mover advantage, but don't expect to gain much more. Don't get trapped putting too much effort into this issue.

Be careful not to confuse something nobody opposes for something everyone supports. I once worked for an organization that wanted to put an initiative on the ballot to lower taxes across the board. In their mind, this was a surefire way to get tax relief passed, and they believed this issue was accessible ground. When they spoke to voters about it, the people always voiced their agreement, but when it came time for action, it was hard to get volunteers excited or to gain widespread support. They did not manage to get their measure on the ballot because they did not correctly understand the type of ground they were standing on.

Ground which can be abandoned but is hard to re-occupy is called entangling. From a position of this sort, if the enemy is unprepared, you may sally forth and defeat him. But if the enemy is prepared for your coming, and you fail to defeat him, then, return being impossible, disaster will ensue.

Entangling Ground: Once you make a statement, you cannot take it back. If you have not yet declared your stance before your opponent speaks out on this issue in a way you had not expected, and you respond to them before doing your due diligence, you may find yourself on the wrong side of an issue with no way to return to neutral ground.

Failing to align yourself with the community leaders in your district loses you the opportunity to outreach meaningfully to the voters within their sphere of influence, and as a result, loses you access to their donors and volunteers. Before making any decision regarding an issue that might be entangling ground, discuss it with your leadership committee. Ensure that it will bring you closer to your allies and not alienate them.

If your opponent missteps on entangling ground, and as a result finds themselves at odds with an important community, capitalize on this to strengthen your own alliances and resources. In the 2016 Presidential Election, Hillary Clinton referred to Trump's supporters as "deplorables." With this, the cat was out of the bag, and it could not be put back in—and as a result many swing and right-of-center voters felt personally attacked. There might have been a way to alienate them from Trump, but instead their allegiance was solidified. This is the danger of entangling ground.

When the position is such that neither side will gain by making the first move, it is called temporizing ground. In a position of this sort, even though the enemy should offer us an attractive bait, it will be advisable not to stir forth, but rather to retreat, thus enticing the enemy in his turn; then, when part of his army has come out, we may deliver our attack with advantage.

Temporizing Ground: This could be something so controversial that neither side wishes to touch it with a ten-foot pole, or so weird and niche that neither side can gain an advantage from taking a position. Stay far away from this issue, and do your best to bait your opponent into making their statement first.

The list of issues that should be important to everyone, but sadly are not, is much longer than any of us would care to admit. There are plenty of good causes in need of a champion, and as soon as you declare your candidacy their supporters will come out of the woodwork. Some hold extreme and less popular views; but of course, they believe their causes are good ones. Because of this, it can sometimes be difficult to distinguish which you are dealing with, so do your research first.

If an issue has few supporters in your district, then it should not be a central part of your campaign. Unless one of your goals is to raise awareness of this issue, you will not be accomplishing your goals by spending your precious little time talking about it.

With regard to narrow passes, if you can occupy them first, let them be strongly garrisoned and await the advent of the enemy. Should the army forestall you in occupying a pass, do not go after him if the pass is fully garrisoned.

With regard to precipitous heights, if you are beforehand with your adversary, you should occupy the raised and sunny spots, and there wait for him to come up. If the enemy has occupied them before you, do not follow him, but retreat and try to entice him away.

Narrow Passes & Precipitous Heights: When one campaign declares itself in support of a position and snatches up all the volunteers and donors too early for the opposition to try to take them, this is an unassailable position. This happens more often in primary elections, or when a candidate holds an opinion that his party is not known for. Supporters will be with your opponent by default as they were the first to occupy the ground. Voters don't like changing horses midstream; this is the first mover's advantage.

Consider being a pro-choice Republican or a pro-life Democrat in a sympathetic district. They may be able to peel a few voters away from their opponent because of their unusual stance, but there exists no compare and contrast material here, and thus no advantage to be gained in a general election. Holding this position may be a requirement to be elected in your district, but know that your opponent will have their support by default and that it will work against you regardless. Do not waste time overcompensating for your inherent disadvantage here.

If you are situated at a great distance from the enemy, and the strength of the two armies is equal, it is not easy to provoke a battle, and fighting will be to your disadvantage.

Positions at a great distance from the enemy: Party-affiliated issues are largely defined by the party you belong to, and are usually at great odds with your opponent's views on the same. There is little chance of actually winning a fight on this issue.

If your district is split fairly evenly and a small group of supporters centers around each position, you will not be able to do much to win an advantage here. But if your opponent is the first to say something about it, there's a good chance your base will think it inflammatory. You can use his words to excite your supporters and recruit extra volunteers. Here the first mover is at a disadvantage.

You may also find yourself in a situation where you believe you can gain an advantage, but you cannot be seen too close to the source. For example, if you discover some disruptive piece of gossip about your opponent, you may not wish to be seen as the one who exposes it, or else it might be viewed as lies and slander. Better to let a third party dangle the bait so that you can respond after the fact. More on scandals is written in Chapter 12: "Attack by Fire".

The natural formation of the country is the soldier's best ally; but a power of estimating the adversary, of controlling the forces of victory, and of shrewdly calculating difficulties, dangers and distances, constitutes the test of a great general.

He who knows these things, and in fighting puts his knowledge into practice, will win his battles. He who knows them not, nor practices them, will surely be defeated.

The circumstances that already exist in the district, before you have even begun your campaign, are where you can find the easiest and often the best advantages over your opponent. If you are aware of a truth not apparent to your opponent, you may trick them into giving the advantage up to you. Victory is found in understanding how to identify these situations, estimating your opponent, and in good planning. Practice on campaigns, and gain whatever experience you can.

If fighting is sure to result in victory, then you must fight, even though the ruler forbid it; if fighting will not result in victory, then you must not fight even at the ruler's bidding.

This advice is aimed at both the candidate and the campaign manager. Moments will arise wherein a staffer has an understanding of the reality of a situation that the candidate lacks. If you, as a campaign manager, find yourself in this position, you must do what you know is best because you are the one with the experience. If the candidate presses you to take a bad stance on an issue, or to use the organization in a disadvantageous manner, you must refuse. This will not be a popular decision, but be secure in the knowledge that it is the right one. This may cause a fight, and it may lose you your job, but if the alternative is piloting a hopeless race in a way you do not believe is right, you will be better off elsewhere. Do not take this to mean that you cannot agree to run a campaign you know will lose, as oftentimes there are mitigating circumstances that might make this race still a good decision.

I regularly have to tell my candidates that what they want to achieve is not possible, the voters in their district do not support that issue, or that they should choose a different district in which to run. I have also agreed to work on campaigns I knew were destined to fail, but I did so because I believed those campaigns were necessary for some other reason, as earlier in this book I have discussed. This might seem contradictory with the earlier advice against allowing insubordination, but together they speak to a balance of power and responsibility; An important dynamic between the manager and the candidate.

If we know that our own men are in a condition to attack, but are unaware that the enemy is not open to attack, we have gone only halfway towards victory.

If we know that the enemy is open to attack, but are unaware that our own men are not in a condition to attack, we have gone only halfway towards victory.

If we know that the enemy is open, and that our men are in a condition to attack, but are unaware that the nature of the ground makes fighting impracticable, we have still gone only halfway towards victory.

Hence the saying: If you know the enemy and know yourself, your victory will not stand in doubt; if you know Heaven and know Earth, you may make your victory complete.

If you know your volunteers are trained and ready but don't know if an incumbent has the organization to defend his seat, you're only halfway to victory. If you know an incumbent is weak or a seat is open, but do not know how strong of an organization you could build, you have only gone halfway to victory. If you know the incumbent is weak and that your organization is strong, but have not yet analyzed the district to see if you can even win, then you have still only gone halfway to victory.

In each campaign, we do our best to lay out our plans, build our organization strategically, and build up our potential energy so that it can be released at the right time toward the right goals. But if we have a poor understanding of the state of our organization, the quality of our plan, the reality of our district, the strength of our opponent, or any of the dozen other considerations needed to win our election, then we will have no chance. Sun Tzu calls this going only halfway toward victory, and as we all know, close only counts in horseshoes and hand grenades.

XI. The Nine Situations

The art of war recognizes nine varieties of ground:
(1) Dispersive ground; (2) facile ground; (3) contentious ground; (4) open ground; (5) ground of intersecting highways; (6) serious ground; (7) difficult ground; (8) hemmed-in ground; (9) desperate ground.

Continuing the concepts of Earth from Chapter 1 and Terrain from Chapter 10, we have the nine types of Ground: A list of situations you may find yourself in while on the campaign trail. As you read through the list consider how the circumstances of your district might place you in this position, and how you'd respond.

Remember that these are best practices, and within every rule, an exception exists. Just be wary of being too ready to believe that your situation is one of the exceptions. More often than not, if you find yourself in a position described above, your best bet is to follow this advice; they are called best practices for a reason.

When a chieftain is fighting in his own territory, it is dispersive ground. On dispersive ground, fight not. I would inspire my men with unity of purpose.

Dispersive Ground: This is any situation happening in your home district and while surrounded by your natural allies; when you are debating an issue you are intimately familiar with, talking to the voters, volunteers, and donors who make up your base of support, or walking door-to-door in the neighborhood where you live. Whenever you engage here, you will have such a large natural advantage that you practically cannot lose. If you're a teacher being attacked on education, if you're an incumbent talking to your donors, etc. It may be true that you cannot be hurt here, but there is also little or nothing to be gained here, so do not fight. Instead, use the opportunity to inspire your supporters and recruit new volunteers. If your opponent attacks you here, remind yourself you are safe.

A few years ago, I worked on a campaign where the local Democrat leaders had started a PAC, and given it a name that sounded like it was a local conservative movement. They raised money from their base of donors and used that money to send mailers to Republican households attacking their Republican opponents for not being conservative enough. For my part, I counseled my candidates to ignore their attacks, to think of what their opponents wanted them to do, and not to change their plans to fight here. This was bait designed to force us off message, we did not respond, and it made no impact on the race.

When he has penetrated into hostile territory, but to no great distance, it is facile ground. On facile ground, halt not. I would see that there is close connection between all parts of my army.

Facile Ground: The edge of the enemy territory. Think of a coalition that has only recently been added to your opponents' list of allies, or voters in a neighborhood that leans slightly to the opposing party. This is probably not where you should be spending your time, and unless your math shows that you need to win these voters to your side, or you have some other good reason for engaging your opponent here, then any victory you do manage to achieve will not be worth the effort. Make sure you have clear communication with all parts of your campaign so that if your volunteers are knocking in this neighborhood or engaging on social media on this topic, you can easily redirect them.

Candidates get distracted easily. They often have the bad habit of watching what their opponent says and does a little too closely. A little of this, under the guise of watching for gaffes or opportunities to attack, can be quite prudent. But more often than not, when the candidate themselves is taking part in it, it tends to lean a little more towards obsession than prudence. As with driving, where your car will slowly tend toward where you're looking, if you're paying too much attention to what your opponent is doing, you are likely to end up reactive rather than proactive. If your opponent gains some ground or adds an important ally, you might find yourself getting jealous, and thinking to yourself that if only they knew the truth about him, they would drop his campaign and sign on with you.

Ground the possession of which imports great advantage to either side, is contentious ground. On contentious ground, attack not. On contentious ground, I would hurry up my rear.

Contentious Ground: This is an asset of great value to both sides; such as the issue that is key to winning the support of voters within your district, or perhaps a well-positioned/wealthy neighborhood of swing voters. If you control this territory, do not be lured out of your defensive position. Your opponent will try to bait you off to attack something else. Bring all resources to bear in defending this ground. Because of its importance to the success or failure of any campaign in your district, if you are in control of the contentious ground, your opponent will be practically obligated to contest you here. Pay special attention to any vulnerabilities you might have, because now is the time for your opponent to strike.

If, on the other hand, your opponent is in control, you should not attack him here. But why not? Certainly, if the issue is important to voters, you will need to wrest control of it to win, right? The point here is that your opponent has the upper hand. The more you try to brute force the issue, the more supporters your opponent will gather.

The biggest mistake that can be made here is believing you are in control of contentious ground when you are not. I advise every candidate to create a Venn diagram of the issues they care about, and the opinions of their district. To populate this chart, polling and other non-biased sources of information gathering are of vital importance. Many people hold the opinion that they are moderate and resemble the average voter in the district. But, if you are a candidate, you can be fairly sure that this is not true. You belong to a party, and if you have views that are on one or other extreme, you are very likely to have friends who support and echo your opinions.

Imagine you are pro-life or pro-choice, and your opponent is the opposite. Each of you is likely to be surrounded by others who share your values. Each of you is likely to believe the district agrees with you, and that this issue is vital to winning the race. If you haven't done any polling, how can you know for sure? Before you dig in your heels, be sure you know that a majority of voters believe the issue is important, and that you hold the upper hand. If you do have the advantage, build up your base of support, and don't spend time fighting your opponent.

Ground on which each side has liberty of movement is open ground. On open ground, do not try to block the enemy's way. On open ground, I would keep a vigilant eye on my defenses.

Open Ground: The opposite of contentious ground, this is an issue of little importance in the district or an area populated mostly by non-voters. There is almost no reason to ever engage here. If your opponent wants to spend their time here, let them. Be wary of being baited into useless actions, and make sure all parts of your campaign know to leave this alone.

Continuing the Choice/Life thought exercise from above, this time imagine that you are running for something like Sheriff or Comptroller, or perhaps running in a district where abortion does not poll amongst the top issues with swing and persuadable voters in your district. What then? The answer is simple: Leave it alone.

I once advised a candidate for Sheriff who made this a central part of his campaign. Despite telling him it was not a relevant issue for a Sheriff candidate to champion, he insisted that he "would not back down"—He believed his cause was noble. But there are only so many hours in the day, and if you focus your attention where you cannot make the most impact, you're willingly and knowingly giving away advantages to your opponents.

> *Ground which forms the key to three contiguous states, so that he who occupies it first has most of the Empire at his command; when there are means of communication on all four sides, it is a ground of intersecting highways. On the ground of intersecting highways, join hands with your allies.*

Ground of Intersecting Highways: Sometimes a single issue is at the root of many smaller ones, or a neighborhood is at the center of the district and is well connected to the rest. In a large district, this may literally be where highways intersect. Use this as a base for building alliances, or maintaining the ones you have. From this position, you can usually build several coalitions. If you have a physical office situated here, you can send staff all over the district. If you win the support of a centrally located neighborhood, your newly recruited volunteers will not have far to travel, no matter where you send them. This is the best position for fundraising.

Consider your economy of action. As said before, there are only so many hours in a day, and only so many days before the election. You are so limited, that if you don't constantly and consistently do what is most efficient you run the risk of falling behind your opponent, who you should always assume is not making the same mistake. Think about what one activity you can do which will advance your position on multiple fronts. Think of the issues that are important to multiple demographics, and which will allow you to build up multiple allies.

Again, remember Tip O'Neill's phrase: *"All politics is local"*. This is a large part of the reason why. When airport noise makes a large impact across your district, a new shopping center or low-income housing development is cropping up in one part of town, or something else which transcends partisanship is happening in your district, it is likely to affect everyone. If it is important to your district, and your party does not have a default stance on this issue, this is likely to be the ground of intersecting highways.

When an army has penetrated into the heart of a hostile country, leaving a number of fortified cities in its rear, it is serious ground. On serious ground, gather in plunder. I would try to ensure a continuous stream of supplies.

Serious Ground: This is the heart of enemy territory, where, by plan or luck, you have been successful; A demographic that is usually a bastion of the opposing party but for some reason supports you, or an issue your opponent should have strong control of, but which has recently leaned in your favor. Sun Tzu uses the word plunder here, and that is exactly what you should be doing. The voters you have just won are very unlikely to already be on your volunteers list, or be counted amongst your donors. In fact, your opponent was probably counting on them for his campaign. Take what you can.

Aim to do as much damage as possible here. If you can dissuade your opponents' base of support from coming out to vote, you can put a serious damper on their organization. If you do fight here, you can count on your supporters to give it everything. No doubt you have in mind some group of his supporters you can chip away at. Be wary of purposefully targeting these people; because unless you have some secret genius strategy, you're not likely to be very successful. But, if you do find yourself successfully, pulling off some amazing strategy, and you manage to peel off some of your opponent's base, you should be prepared to capitalize on it.

Mountain forests, rugged steeps, marshes and fens—all country that is hard to traverse: this is difficult ground. On difficult ground, keep steadily on the march.

Difficult Ground: Difficult says it all. This is a confusing issue or a neighborhood that is difficult to walk in. If you are cornered about your position on a difficult issue, try to pivot off of the subject and move on as quickly as possible—avoid saying anything before you've done your research. In the field, you should pass difficult neighborhoods by when your team is out walking door-to-door. Where the voters are not easily accessible in-person other options such as phone, text, email, and good old-fashioned snail mail are all better suited.

I have worked for a handful of conservative candidates who were supportive of gun rights. This was usually a popular issue in their districts, but after a few high-profile public shootings, it put them in quite a precarious situation. How to remain supportive, signal to our base that our position would not waver, and still be respectful of the tragedy? Proverbs remind us that even a fool, when he holdeth his peace, is counted wise. "I'll have to look into it." or "I still need to do my research." are usually perfectly respectable answers. You can often avoid gaffes by keeping your mouth shut. Engage only when you are prepared.

Ground which is reached through narrow gorges, and from which we can only retire by tortuous paths, so that a small number of the enemy would suffice to crush a large body of our men: this is hemmed in ground. On hemmed-in ground, resort to stratagem.

Hemmed-in Ground: Any situation that is challenging to navigate and potentially perilous. This could be analogous to a change in your well-known stance on a controversial issue, or an attempt to conduct in-person outreach in a remote area of the district. Undertakings like these are likely to be costly, inefficient, and have a low probability of success.

Your most effective strategy in these circumstances is adaptability and quick thinking. For instance, if you find yourself far from your office without sufficient campaign literature, or if too few volunteers turn up at an event, retreating is not an option. Instead, think on your feet and transform the situation into a different type of event. If you have taken an unpopular stance on an important issue and need to change course, strategize to ensure your allies stand by you during the transition.

On the subject of Hemmed In Ground, Sun Tzu advises us:

> *The further you penetrate into a country, the greater will be the solidarity of your troops. Make forays in fertile country to supply your army with food. Carefully study the well-being of your men, and do not overtax them. Concentrate your energy and hoard your strength. Keep your army continually on the move, and devise unfathomable plans.*

If you are an invading force, such as a candidate venturing into a new district for the first time or attempting to sway a new group of voters by changing your stance on a hot-button issue, the ability to maintain and even strengthen your team's unity becomes all the more important.

Changing stance on a controversial issue can potentially cause discord within your team, especially if members strongly align with the previous stance. However, the unity of your team can still be maintained if the change is managed carefully. These are the narrow gorges that Sun Tzu spoke about. Refer back to Chapter 5: "Energy", which discusses strategic team building; a tightly-knit team will not only be more productive but will also be more successful. When they see their efforts contributing to a potential victory, their morale will soar.

But remember Sun Tzu's caution against overtaxing your troops. Preserve your team's energy for the crucial days leading up to the election. Invest time in team-building and organizational training. Keep your team growing and motivated until the final moment, and do your utmost to prevent your opponent from discovering your plans.

Ground on which we can only be saved from destruction by fighting without delay, is desperate ground. When there is no place of refuge at all, it is desperate ground. On desperate ground, fight. On desperate ground, I would proclaim to my soldiers the hopelessness of saving their lives.

Desperate Ground: A position which you can only defend by fighting. When you're pressed to make a statement on a hot-button issue by an important community leader in your district, you can't simply pivot away. If this person's support is critical to your victory, or when you arrive to knock on doors in a target neighborhood and find that your opponent has already been here talking with voters, you must fight. Throw all your power into winning this battle as the rewards likely far outweigh the cost.

Every campaign will have a few of these. Independent, and persuadable voters are often targeted by both sides, and winning their support as early as possible goes a long way toward victory on election day. Every high-propensity independent voter you can win to your side is 2 votes in your pocket; one more vote for you, and one less vote for your opponent.

Imagine your campaign is hit by a major scandal or negative news story that threatens to derail your chances, or suppose you've released a policy proposal that is met with significant backlash from a demographic or stakeholder group that was critical to your success. Possibly you will find yourself in a high-stakes debate or public forum, and you suddenly find yourself on the back foot due to a strong attack from your opponent. These are desperate situations, and they may call for drastic measures.

Sun Tzu advises us on how to handle your team when faced with Desperate Ground:

> Throw your soldiers into positions whence there is no escape, and they will prefer death to flight. Officers and men alike will put forth their uttermost strength. Soldiers when in desperate straits lose the sense of fear.
>
> Thus, without waiting to be marshaled, the soldiers will be constantly on the qui vive; without waiting to be asked, they will do your will; without restrictions, they will be faithful; without giving orders, they can be trusted.

Your lives may not be in danger, but in many cases, the future of your ideals is. Make sure your entire team understands that without action, the life they love may die. They must understand this completely; regardless of whether it is true or not. It would be irresponsible of me to suggest you should openly lie to your staff, but, there are often multiple possible outcomes. It would be advantageous if they felt their values and ideals are threatened. Certainly, your opponent will be trying to nurture the same attitudes within his supporters.

In the heat of a campaign, every member of your team needs to feel that they're on that desperate ground, where their utmost effort is the only thing that can secure victory. This sense of urgency and shared responsibility can turn a disparate group of individuals into a cohesive, unstoppable force. Each phone call made, each door knocked, and each donation given becomes a crucial step towards victory. In these moments, the campaign is not just about the candidate, but about each person who is contributing their time, energy, and resources to the cause. It's their fight too, and they'll be more likely to put forth their maximum effort when they understand that they're part of something bigger than themselves.

If your volunteers believe that the only thing between your victory and a crushing defeat is their hard work, they will be incentivized to work harder. Remind them of what is at stake, what happens if you lose, and what can be gained by victory. With this knowledge driving them, your volunteers will not need to be harassed to show up to a walk-day or a phone bank, your donors will not need to be constantly reminded to give, and your staff will show up every day to work for their win bonus.

* * *

Those who were called skillful leaders of old knew how to drive a wedge between the enemy's front and rear; to prevent co-operation between his large and small divisions; to hinder the good troops from rescuing the bad, the officers from rallying their men.

Do what you can to improve your position and try to press your advantage when you find your opponent in a disadvantageous position. If you can do anything to separate him from his supporters or disrupt his efforts, and doing so would not bring you a greater disadvantage, you should probably do it. On the other side of that same coin, beware of your opponent finding out you're compromised, because they will also be following this advice.

If asked how to cope with a great host of the enemy in orderly array and on the point of marching to the attack, I should say: "Begin by seizing something which your opponent holds dear; then he will be amenable to your will."

If your enemy is in a good spot, and his campaign is humming away like a well-oiled machine, your best bet is to use bait. Attack something he must defend, and he will hurry to defend it. I have cautioned above that you should not be too zealous in believing you've found the ideal strategy to chip away at your opponent's base. But if you are going to lose unless something drastic happens, then you are going to need to make something drastic happen, and this is the way to do it.

Rapidity is the essence of war: take advantage of the enemy's unreadiness, make your way by unexpected routes, and attack unguarded spots.

As Sun Tzu says, "Rapidity is the essence of war". Once you have identified a situation or circumstance that will be beneficial to you, or harmful to your opponent, you have a duty to press your advantage immediately. Once you've won (or lost) an individual battle, it's time to begin the next one. Attack relentlessly, work tirelessly, and adjust your plan constantly.

The different measures suited to the nine varieties of ground; the expediency of aggressive or defensive tactics; and the fundamental laws of human nature: these are things that must most certainly be studied.

Success in warfare is gained by carefully accommodating ourselves to the enemy's purpose. By persistently hanging on the enemy's flank, we shall succeed in the long run in killing the commander-in-chief. This is called ability to accomplish a thing by sheer cunning.

In the campaign world, we don't focus enough effort on the time between election cycles. The political parties, their consultants, and other career staffers who have seen many campaigns tend to be fixated on 2-year or 4-year cycles because that is how often campaigns happen. After an election, we take vacations, then get back to recruiting candidates, finding new jobs, and having sales calls with new prospective clients. But if you take your time, and build an organization with longevity of purpose, (such as supporting an issue that is important to the district, or registering new voters from your supportive demographic), you can begin your campaign long before the election starts. Eventually, you will be "recruited" to run for office; it'll be quicker than you think before the party comes calling.

On the day that you take up your command, block the frontier passes, destroy the official tallies, and stop the passage of all emissaries. Be stern in the council-chamber, so that you may control the situation.

If the enemy leaves a door open, you must rush in. Forestall your opponent by seizing what he holds dear, and subtly contrive to time his arrival on the ground. Walk in the path defined by rule, and accommodate yourself to the enemy until you can fight a decisive battle.

At first, then, exhibit the coyness of a maiden, until the enemy gives you an opening; afterwards emulate the rapidity of a running hare, and it will be too late for the enemy to oppose you.

With an organization built and ready to pivot, with the research on your district and opponent completed, and a solid but flexible campaign plan, you'll be ready to strike when your targeted district is open to you. As he said in Chapter 2: "Waging War", thus, though we have heard of stupid haste in war, cleverness has never been seen associated with long delays.

Part 6: Scandal & Spies

"An investment in knowledge pays the best interest. The only thing more expensive than education is ignorance."

Benjamin Franklin

In the context of campaigns, this means understanding the full array of strategies and tactics at your disposal—even those that may seem contentious. Part 6: "Aggression & Intelligence," navigates the often turbulent waters of scandals, negative campaigning, and the use of spies to gather intelligence.

In Chapter 12: "The Attack by Fire", we'll explore how to harness the power of information, timing its release for maximum impact while mitigating potential fallout. Chapter 13: "The Use of Spies." underscores the importance of research and intelligence gathering in campaigns.

While some might find these topics distasteful, the fact is that they are aspects of modern campaigning. Whether you choose to employ these tactics or not, understanding them will allow you to better defend against them should your opponent choose to be less honorable than you. In this part of your journey, may you find the wisdom to navigate these complex situations with integrity and foresight.

XII. The Attack by Fire

There are five ways of attacking with fire. The first is to burn soldiers in their camp; the second is to burn stores; the third is to burn baggage trains; the fourth is to burn arsenals and magazines; the fifth is to hurl dropping fire amongst the enemy. In order to carry out an attack, we must have means available.

I think of fire in this context as an allegory for scandals. It kills campaigns without regard; it spreads without outside influence; and if you try to harness it, you can easily get burned instead unless you're very careful. A strong enough scandal can alienate volunteers, force you to use up your budget to quench it, and will dry up your stream of donations in a flash.

To effectively use scandal as a weapon, you must have something to use. It would be pointing out the obvious to say you can't just make something up, but I commonly see candidates internalizing their own hatred and exaggerating some small thing about their opponent until they believe it is scandal worthy—this is essentially the same thing as making it up.

Do your research, and discover the things your opponent doesn't want you to know. If you're lucky enough to find something, hold on to it until the time is right. If you are too early people will have time to forgive and forget, and if you are too late people will not have the chance to hear about it before they vote. More and more voters are moving towards absentee and early voting so the sweet spot is shrinking. Time your October surprise well, and give the flames time to grow before early voting begins.

In attacking with fire, one should be prepared to meet five possible developments:

> *(1) When fire breaks out inside the enemy's camp, respond at once with an attack from without.*

When a scandal breaks out that targets your opponent, it is time to strike; call him out publicly, hold his feet to the fire, and do as much damage to his coalitions and supporter base as possible.

> *(2) If there is an outbreak of fire, but the enemy's soldiers remain quiet, bide your time and do not attack.*

In this regard, acting too early is almost as bad as acting too late. Remember that your perception of events is going to be skewed in the opposite direction of your opponents. As such, don't become overzealous or let your anxiety force you into action. If his supporters are not up in arms, then perhaps the fire has not reached its full potential, or perhaps he got to it in time. Either way, the time is not yet right.

>   *(3) When the force of the flames has reached its height, follow it up with an attack, if that is practicable.*
>   *(4) If it is possible to make an assault with fire from without, do not wait for it to break out within, but deliver your attack at a favorable moment.*

When you know the time is right, and you see an opportunity to use what you have, you must strike while the iron is hot, as they say. Do not hesitate or you may let the opportunity pass you by, and you will likely not get another shot at it. Use what you have when he is least expecting it, and when it will do the most damage.

>   *(5) When you start a fire, be windward of it. Do not attack from the leeward. A wind that rises in the daytime lasts long, but a night breeze soon falls.*

As mentioned above, fire is dangerous for all those involved, and sometimes an issue is one you should not touch with a 10-ft pole. Perhaps you hold the same view, though it may not be as acceptable to your base. If you and your opponent share the same views on a subject, and the voters in your district are against it, if you show your hand you're going to get burned too. Consider again the example of a pro-choice Republican or a pro-life Democrat, and how silly it would be for them to attack their opponent across the aisle. Instead you might tap one of your allies who is better positioned to attack him. Remember to be careful that when instigating conflict in your opponent's camp, you do not get burned yourself.

> *In every army, the five developments connected with fire must be known, the movements of the stars calculated, and a watch kept for the proper days.*

If you want to succeed in deceiving your opponent and causing damage to his base, you should be aware of all the possibilities. Outline as many outcomes as you can, and plan contingencies for all possible developments. Do your research, and be patient for the time to strike. Do not go on the offensive too early, and don't miss your opportunity or it will be gone forever. Again, be wary that your perspective might skew the reality of the situation. So rely on the intelligence you've gained and the opinions of unbiased advisors.

As in all parts of the campaign, beware of your opponent putting you in the same situation. If scandal sets fire to your camp, it will be your reputation that is under attack, your good name being dragged through the mud. Do not fight just to satisfy your ego. Fight only when it is advantageous to do so.

Once you've successfully disrupted your opponents' plans, and the loyalty of his supporters wanes, you will have succeeded in seizing the advantage and moved one step closer to defeating him. An incumbent who is forced out is rarely successful at being elected again in the future.

XIII. The Use of Spies

Hostile armies may face each other for years, striving for the victory which is decided in a single day. This being so, to remain in ignorance of the enemy's condition because one grudges the outlay of a hundred ounces of silver, is the height of inhumanity. One who acts thus is no leader of men, no present help to his sovereign, no master of victory.

Thus, what enables the wise sovereign and the good general to strike and conquer, is foreknowledge. Now this cannot be elicited from spirits, nor by any deductive calculation. Knowledge of the enemy's dispositions can only be obtained from other men.

Campaigns get more expensive every cycle. We are constantly hearing about some contentious congressional race that is the most costly in history. Despite this, most large campaigns employ a research director whose sole duty is to keep tabs on your opponent and his allies. This might seem like an extravagant expense, but it's nothing compared to the cost of a single mailer, or a week with your whole team in the field. The entirety of your years-long campaign will come down to a single election. Make sure you know everything there is to know.

You will work for years for a result that is decided in a single day. This is the reason that using spies is so important. Don't skimp!

* * *

Hence the use of spies, of whom there are five classes:
(1) Local spies; (2) inward spies; (3) converted spies;
(4) doomed spies; and (5) surviving spies.

When these five kinds of spy are all at work, none can discover the secret system.

Intelligence gathering functions as your campaign's early warning system, equipping you with crucial foresight. Echoing the 14th Law from "The 48 Laws of Power," some of the most valuable insights can come directly from your own observations and interactions. In crafting your intelligence strategy, a nuanced blend of the five types of spies should be strategically deployed, just as every other element of your campaign requires careful planning

Having local spies means employing the services of the inhabitants of a district.

Local Spies: These are simply the inhabitants of your district. Elsewhere referred to as local guides, these are the folks that have their ear to the ground. They should be as unbiased in their opinions as possible and should be well-connected.

Make sure a few of this type of spy are on your advisory team early. As written about before, hyper-local issues are often the key to winning the district, and you will need someone to make you aware of them.

Having inward spies, making use of officials of the enemy.

Inward Spies: You can also profit off the staff and volunteers of your opponents. If you sign up for their email newsletter or send an email from a fake account asking to intern for them, your opponents will be more than happy to tell you themselves about the state of their organization. You can usually find a list of the events they have coming up on their website or social media pages.

Take this a step further, and move beyond just the official channels the campaign uses to communicate. Research his staffers and advisors, then follow their social media accounts. They will tell you their thoughts and from this, you can ascertain their plans.

The enemy's spies who have come to spy on us must be sought out, tempted with bribes, led away, and comfortably housed. Thus they will become converted spies and available for our service.

The end and aim of spying in all its five varieties is knowledge of the enemy; and this knowledge can only be derived, in the first instance, from the converted spy. Hence it is essential that the converted spy be treated with the utmost liberality.

Converted Spies: This is the inverse of the Inward spies. If you find your opponent is subscribed to your newsletter, you can send a special email just to him.

In primaries, or when an opponent has pissed off a certain coalition, his volunteers may change camps, and they make the most successful spies. If the fire of scandal has burned him recently, his previous allies are likely to dump his dirt for all to see.

The most is written about converted spies because they are the best source of information coming directly from your opponent. If someone who has seen your opponent's campaign plan is driven away from his camp, you want to discover what he knows.

Having doomed spies, doing certain things openly for purposes of deception, and allowing our spies to know of them and report them to the enemy. It is owing to his information, again, that we can cause the doomed spy to carry false tidings to the enemy.

Doomed Spies: An intern, volunteer, or staff of your opponents' who attends your events. This is extremely common in high-profile races. The more trusted interns and volunteers a campaign has, the more likely some will want to participate in this sort of activity. If you can identify them when they come to your fundraisers or volunteer days, you can make sure they see only what you'd like them to see. Intercept them with your own spies, and make sure the information they bring back to your opponent is misleading and wrong.

Surviving spies, finally, are those who bring back news from the enemy's camp. Lastly, it is by his information that the surviving spy can be used on appointed occasions.

Surviving Spies: Just like doomed spies, except these are your interns, volunteers, or staff, who attend your opponents' events to bring back their materials, information about their organization, and the details of their plans. Make sure they were not intercepted and this information will be some of the most valuable.

Hence it is that which none in the whole army are more intimate relations to be maintained than with spies. None should be more liberally rewarded. In no other business should greater secrecy be preserved.

Spies cannot be usefully employed without a certain intuitive sagacity. They cannot be properly managed without benevolence and straightforwardness. Without subtle ingenuity of mind, one cannot make certain of the truth of their reports.

Simply put, research is central to almost everything you will do in your campaign. Knowing your enemy will come down to how closely you watch him. So watch him closely. Be as subtle as you can be and instruct your staff and volunteers to do the same, but use spies as often as possible.

If a secret piece of news is divulged by a spy before the time is ripe, he must be put to death together with the man to whom the secret was told.

Loose lips sink ships, and some people just cannot keep their mouth shut. If a staffer divulges information he should be fired. If a volunteer leaks your plans to your opponent, restrict his access to all but the basic info. This type of person tends to be a repeat offender.

Whether the object be to crush an army, to storm a city, or to assassinate an individual, it is always necessary to begin by finding out the names of the attendants, the aides-de-camp, and door-keepers and sentries of the general in command. Our spies must be commissioned to ascertain these.

Hence it is only the enlightened ruler and the wise general who will use the highest intelligence of the army for purposes of spying and thereby they achieve great results. Spies are a most important element in war, because on them depends an army's ability to move.

As was said in Chapter 3, "*If you know the enemy and know yourself, you need not fear the result of a hundred battles.*" The most effective way to know your enemy is through the strategic use of spies. Properly using each of the five types of spies can give you a more complete understanding of your opponent's strategies, strengths, and weaknesses. As with all other parts of this book, remember that your opponent will be using some or all of these tactics against you, so for your part, do your best to stop them from being successful.

Final Thoughts

"The whole state must be so well organized that every Whig can be brought to the polls. So, divide the county into small districts and appoint in each a committee. Make a perfect list of the voters and ascertain with certainty for whom they will vote. Keep a constant watch on the doubtful voters and have them talked to by those in whom they have the most confidence. On Election Day see that every Whig is brought to the polls."

Abraham Lincoln, Campaign Circular from January 1840

This quote is one of my favorite political quotes. Despite coming from nearly 200 years ago, what Lincoln describes here is almost exactly the way we currently conduct our campaigns: we cut our districts into smaller sections, recruit captains in each, perform a thorough voter identification effort, and try to sway the undecideds with messages from respected neighbors and leaders in the community. Finally, with the start of Absentee and Early voting, we kick off our GOTV efforts to make sure every last supporter shows up to the polls.

Candidates often fall into the trap of believing their message is so powerful, that their name is so well known, or that their strategy is so clever that their campaign does not need to follow the traditional path. While at points in this book, I have encouraged trying to find a way to win without fighting, this is seldom possible. As said by Carl Von Clausewitz in his treatise On War, *"War in its literal meaning is fighting."* and you would do well to remember that.

So, focus your efforts on crafting a good strategy, out-maneuvering your opponent, or working smarter instead of harder, but in the end be prepared to conduct your campaign in the traditional way. You will still need to talk to voters, communicate your message via the method by which they prefer to be contacted, develop real relationships with as many of them as possible, and ensure that on election day your voters show up to the polls. Indeed, campaigning is hard work and requires commitment, strategy, and perseverance. As you reach out to voters, you will learn more about your community and its needs, forging connections and building relationships that can last a lifetime.

Remember to take a moment to reflect on why you originally embarked on this campaign. It is your passion for service, your belief in a better future, and your commitment to the people that make all the effort worthwhile.

Now, go forth, armed with the wisdom of Sun Tzu and the pragmatism of Lincoln, but also your ideas and values. Connect, communicate, and inspire. No matter what the results are on election night, the very act of running for office shows your dedication to democracy and service. This, in itself, is a victory.

About the Author

Caitlin Huxley is an expert campaign strategist, certified project manager, and history buff with a passion for empowering new candidates to overcome challenges and achieve success in their campaigns.

Before founding Huxley Strategies and developing the Campaign Managers Toolbox, Caitlin served the Illinois State House Republican Organization, the Illinois Opportunity Project, and the Chicago Republican Party. Together, they successfully rebuilt party infrastructure, and elected Republicans to the state legislature.

Today, Huxley Strategies trains, develops, and equips candidates and organizations with data-driven strategies to navigate the political landscape and achieve their goals. Over the past decade, Caitlin's clients have consistently outperformed the average, implementing forward-thinking policies and transforming their states.

Acknowledgments

I would like to extend a huge thank you to everyone who helped to make this book possible.

Firstly, thank you to my wife Jessica for her constant support and encouragement.

Secondly, thank you to Jayme Siemer, my friend and mentor, without whom neither this book nor my career in politics would have been possible.

Finally, thank you to my dad, as well as my friends and colleagues Andy Bakker and Jesus Solario who read this book early on and gave me their thorough reviews.

We need your help! Please Review!

Thank you for reading **Ancient Wisdom for Modern Campaigns**. Your support means a lot to me, and I hope you found it valuable and enjoyable.

If you have a moment, I would greatly appreciate it if you would leave a review on Amazon & Goodreads. Your feedback helps other readers discover my work and help me to improve.

Thank you again,
Caitlin Huxley

Find Free Guides and Downloadable Resources at:

http://www.HuxleyStrategies.com

The Campaign Managers Toolbox is designed to provide new candidates, campaign staff, and volunteers with the skills they need to succeed in the political arena. These guides cover essential topics such as organization building, fundraising, voter contact, GOTV (get out the vote), and more.

Designed for both newcomers and seasoned veterans, these guides and templates will set your campaign on the path to victory and help you navigate the complexities of modern politics.

Caitlin Huxley

www.ingramcontent.com/pod-product-compliance
Lightning Source LLC
Chambersburg PA
CBHW052132270326
41930CB00012B/2851

* 9 7 9 8 9 8 9 0 7 3 7 0 2 *